GREAT NORTHERN ATLANTICS

GREAT NORTHERN ATLANTICS

JAMES S. BALDWIN

First published in Great Britain in 2015 by
Pen & Sword Transport
An imprint of Pen & Sword Books Ltd
47 Church Street
Barnsley
South Yorkshire
S70 2AS

Copyright © James S. Baldwin, 2016

ISBN 978 1 78346 367 1

The right of James S. Baldwin to be identified as the author of this work has been asserted by him in accordance with the Copyright, Designs and Patents Act 1988. All rights reserved. No part of this publication may be reproduced or transmitted in any form or by any means, electronic or mechanical, including photocopy, recording or any information storage and retrieval system, without the prior written permission of the publisher, nor by way of trade or otherwise shall it be lent, re-sold, hired out or otherwise circulated without the publisher's prior consent in any form of binding or cover other than that in which it is published and without a similar condition including this condition being imposed on the subsequent purchaser.

© James S. Baldwin 2016

Typeset by Pen & Sword Books Ltd

Printed and bound by Imago Publishing Limited

Typeset in Palatino

Pen & Sword Books Ltd incorporates the imprints of Pen & Sword Archaeology, Atlas, Aviation, Battleground, Discovery, Family History, History, Maritime, Military, Naval, Politics, Railways, Select, Social History, Transport, True Crime, and Claymore Press, Frontline Books, Leo Cooper, Praetorian Press, Remember When, Seaforth Publishing and Wharncliffe.

For a complete list of Pen and Sword titles please contact
Pen and Sword Books Limited
47 Church Street, Barnsley, South Yorkshire, S70 2AS, England
E-mail: enquiries@pen-and-sword.co.uk
Website: www.pen-and-sword.co.uk

I dedicate this book to my daughter, Chloe, for inspiring me to write it.

CONTENTS

ACKNOWLEDGEMENTS

I would like to thank the many people who have answered my questions, provided images and items of memorabilia for me to examine and who have recalled personal stories of these magnificent machines.

In particular, I would especially like to thank Jonathan Clay, Barry Collins, Michael J. Collins, Antony M. Ford, Richard H.N. Hardy, David Jones, Norman Lee, John Scott-Morgan, L.V. Reason, Roger Sinar, Neville Stead, Ted Talbot and Peter N. Townend.

I also thank my wife, Harriett, for her unfailing encouragement.

James S. Baldwin,
London, 2014

FOREWORD

Peter N. Townend, former District Motive Power Superintendent at King's Cross 1956–1963

Peter N. Townend was born in 1925 and grew up in Doncaster with a fascination for steam and railways, which were all around him, and passed the scholarship to the Grammar School at the age of ten. He has clear memories of the first night that the 'Silver Jubilee' train ran north from King's Cross to Newcastle, hauled by the locomotive *Silver Link*, where it seemed that the whole town had turned out to see the first new high speed train pass through Doncaster, with a long wail on the chime whistle at the unheard speed of 70mph! At sixteen he became a 'premium apprentice' on the LNER, starting in B shop at Doncaster on a screwing machine, making the nuts fitted on drawbars. He obtained his HNC and as a qualified engineer went to the Crimpsall Shops at Doncaster repairing steam locomotives. Upon reaching the age of twenty-one, he requested to work in the drawing office and after three years, he applied to go into the motive power department and was sent to Norwich as a progress man. He then went to Melton Constable to take charge of the depot and spent the next six months looking after the twenty-seven engines allocated there, responsible for the work and staffing arrangements. This also included managing the sub-depots at Norwich City, Cromer and Yarmouth Beach, before being transferred to the Cambridge district. After Cambridge, he worked at Liverpool Street, at the Motive Power Headquarters under L.P. Parker and became a Mechanical Foreman Learner. He was then appointed Shed-Master at Hatfield and after about a year he went to Boston, only to return south to cover Hitchin for a few months. At the age of thirty-one, he was appointed as the District Motive Power Superintendent at King's Cross, where he was responsible for the fitting of the Kylchap blast-pipe and double chimney arrangement with German style smoke deflectors, to aid freer steaming of the A3 class Pacifics. Peter is an acknowledged expert on all things GNR and LNER and has also written numerous articles and books about railways.

Here are Peter's thoughts about the Great Northern Railway's Atlantics.

Engines with Atlantic 4-4-2 wheel arrangements, were a familiar sight to Peter Townend when he was a young boy living in Doncaster. Usually, one stood either on the turntable or nearby at the north end of Doncaster Station, acting as the pilot engine ready to cover any failures. At this time the work of the Atlantics was already in decline, which process was made quicker with the arrival of Pacifics on main line work. Here, we see the driver of large boiler Atlantic 3277 trying to occupy his time as he waits for a job with his pilot engine. (Author's Collection)

The Great Northern Atlantics

The Atlantics were familiar to me as a boy at Doncaster, and usually one stood either on the turntable or nearby at the north end of Doncaster station. This was the pilot to cover any failures, although it did not seem to do very much. The work of the engines was already in decline and pacifics soon replaced the class on main line work.

My personal involvement with the engines was when I was a

Peter Townend's personal involvement with Atlantic type engines started when he was a 'premium apprentice' in the Crimpsall Shops of Doncaster Works. Here is the letter sent to Peter's mother, confirming Peter's successful application. This priceless document has been signed by no other person than A.H. Peppercorn himself, who was the last Chief Mechanical Engineer of the LNER.
(Peter N. Townend Collection)

E. THOMPSON
CHIEF MECHANICAL ENGINEER
A. H. PEPPERCORN
ASST. CHIEF MECHANICAL ENGINEER
TELEPHONE 4417/8
TELEGRAPHIC ADDRESS
"MECHANICAL C/o NORTHEASTERN, DONCASTER"
REFERENCE S.30/27.

JH.

Letters to be addressed
THE CHIEF MECHANICAL ENGINEER
(M. E. OFFICE)
LONDON & NORTH EASTERN RAILWAY
DONCASTER

21st October, 1941.

Mrs. Townend,
"Glencairn"
21, Granby Crescent,
DONCASTER.

Dear Madam,

APPLICATION FOR PREMIUM APPRENTICESHIP FOR SON.

Confirming my telephone message this morning, I am prepared to take your son into these Works on the conditions named. Please let me know when he will start here.

Yours faithfully,

'premium apprentice' in the Crimpsall Shops and worked on Bob Senior's pit for a period during the war. Materials were in short supply and I remember being sent round the scrap yard and anywhere else to find black nuts and bolts to use on repaired locomotives on the pit.

A large-boiler Atlantic was the largest locomotive I worked upon. When completed I was sent on the engine for its trial trip to Barkston and back before being released to the Carr Loco Depot. There was the wonderful aroma of hot black paint and hot oil liberally used on the engine by the weigh-house crew who

This view was taken during Peter Townend's time as a premium apprentice at Doncaster Works. Peter is seen standing immediately beneath the number of small boiler Atlantic 3259, which was originally completed in June 1903. It was one of the last small boiler Atlantics to be withdrawn from service, on 2 October 1943.
(Peter N. Townend Collection)

Here is a view of Lea Road Station, Gainsborough from a previous era. (Author's Collection)

manned the engine on these trips with the odd sizzle from a steam joint. Many Atlantics were now being scrapped but I remember another Atlantic on the next pit receiving new frames, cylinders and boiler only to be scrapped on its next visit to Doncaster.

One of my earliest footplate rides had been on an Atlantic, not the first, however, as that had been on a Caledonian 4-4-0 on the school train at Carstairs. The Atlantic ride had been arranged during the night with a Sheffield crew when I was put into the front corner of the cab to make sure I did not fall off when the engine lurched.

During my period at the Running Sheds in 1946 I went out with the breakdown crane on numerous occasions on both lifting jobs and to derailments, which in those days were very common. On this occasion N° 4403 of New England had split the points in Garden Sidings at the south end and had become derailed, all wheels including the tender. Engines were turned there on the triangle of lines between St James Bridge and Balby Bridge and prepared for their return workings. The rerailment was complicated and took one hour and forty minutes to accomplish using a wire rope, which bent the draw hook, jacks and ramps. I later worked on the engine on the drop pit to replace the trailing wheel helical spring bolts, which had been bent in the derailment.

On another occasion I was sent with a fitter and mate on renumbered Atlantic N° 2869 to rescue a J39 class N° 2706, which had failed, blocking the line about three quarters of a mile south of Gainsborough Lea Road Station while working a train of empty coal wagons. The Atlantic was loaded with a few tools of the heavy variety, leaving Carr Loco at about half past two and arriving on the scene at quarter past four in the afternoon. The right-hand gudgeon pin nut had worked loose on the J39, allowing the pin to work out and catch the motion plate frame stretcher shearing the gudgeon pin half way down the threads, breaking the crosshead slides, bending the piston. The connecting rod was bent and detached at the small end amongst further damage and the engine could not be moved. An attempt was made to take down the connecting rod by removing the big end bolts but this proved to be impossible in a short time. The solution to clear the line was to use a hemp rope through the small end eye and taking it over the boiler hand rail. It was held by me heaving away, lifting the connecting rod to clear the front axle. It took over an hour to slowly draw the J39 into a siding at Lea Road Station where one of the big end bolts was sawn through and the other parts taken down and stowed on the running plate. N° 2869 was used to work the coal empties away and eventually a WD 2-8-0 arrived so that we could return to Doncaster.

My last involvement with an

Large boiler Atlantic 3274 is seen in happier times as it works a prestigious Pullman service. Withdrawn from normal service on 2 May 1946, it then carried out stationary boiler duties at Doncaster Works, from September 1947 and was finally cut in March 1952. (Author's Collection)

Atlantic was much more impressive and came about fortuitously. I was riding on a run from Doncaster on 'Green Arrow' N° 4817 to King's Cross, which was hardly keeping time in 1946 on a main line express at a time when arrivals were generally late. When passing Cadwell there were some bangs and the driver called out 'Keep your head in.' After the train had come to a stand, I crawled underneath the engine and found the middle big end brasses were missing, both the little end oil cup and the cork of the big end were also missing. The marine type of middle big end fitted to the Gresley three-cylinder engines was useful on these occasions, as the engine could be worked forward at very slow speed to Hitchin without any dismantling. At Hitchin Atlantic N° 4404 replaced the V2 and worked the train to King's Cross. After getting the train on the move the 'Atlantic' kept time to London. This was the only ride I had on an Atlantic on the Great Northern main line doing the job it was built for and I was most impressed. The engine was undoubtedly worked hard with the lever reverse near full gear as it was difficult to notch up on the move but the boiler steamed well. Although the engines were now being scrapped they still had a limited role as they could be turned on short turntables such as at Hitchin and Peterborough. The engine was withdrawn the following year.

The Atlantics had worked the GN main line for many years and Gresley had not rushed to replace them, using them on heavy Pullman workings into the 1930s. N° 251 and N° 990 well deserve their place in the National Collection, and how pleasing it is to see a Brighton version being constructed so that one of these engines will one day soon be seen at work again.

Peter N. Townend
Paignton, June 2013

Although ten small boiler Atlantics were built in 1903, the major effort in constructing engines for use on express passenger engines was confined to those of the 'big-brother' version – the large boiler Atlantics, of which ninety-four examples were built between 1902-10. Here is a large boiler Atlantic doing what it was designed to do – working a heavy teak-bodied express passenger service at speed. (Author's Collection)

One of the last three large boiler Atlantics to be constructed at Doncaster Works was 4459, which was completed in November 1910. It is seen here working a Pullman service. It lasted until July 1943. (Author's Collection)

Having successfully completed his premium apprenticeship at Doncaster Works, Peter Townend progressed in his career and became appointed Shed-Master at Hatfield. Then the post of Shed-Master for King's Cross become vacant, although someone else was covering. There were over thirty applicants and he wasn't on the list for interviews, which were held one morning. At lunch time, he received a telephone call to go up and see the Assistant Motive Power Officer, when he was told that he had been appointed as Shed Master at King's Cross. He was under thirty-one years old and he was vaguely told that they wanted somebody to improve the reliability of the locomotive fleet. Here is Peter sitting at his desk at Top Shed, surrounded by a mass of paperwork. (Peter N. Townend Collection)

Having been appointed as Shed Master at King's Cross, Peter Townend set about looking into the various problems which were causing poor reliability and availability of his A3s, A4s and V2s. The depot had considerable difficulty in recruiting staff in the London area as railways were not seen to be attractive to schools. This resulted in engine examinations not being carried out at the appropriate time and repairs were taking longer. Peter asked his superintendent if they could use Doncaster Works to help alleviate the problems. After meeting the Works Manager at Doncaster, who wanted to help in any way he could, it was arranged that all of Peter's Gresley Pacifics would be sent in as they became due for N° 6 'valves and pistons' examinations. This arrangement resulted in his Pacifics running higher mileages between problems. It was noted on the mileage return that six A4s out of the nineteen allocated to King's Cross, which were still at work on the main line, had mileages of over 100,000 miles since their works overhaul, something not previously achieved. This image shows a magnificent line-up at King's Cross depot during Peter's tenure there. (Peter N. Townend)

INTRODUCTION

At the start of the twentieth century, the new fangled Atlantic wheel arrangement, which had originated in the United States of America, had been specifically developed for mainline passenger express services and was the one that every railway company wanted to have and to show off.

The first 4-4-2 type tank engines to be built in this country were constructed between 1880 and 1892 for the London, Tilbury and Southend Railway (LTSR) by Sharp, Stewart & Company and Nasmyth Wilson & Company. The LTSR's 1 class, were numbered 1-36 and had 6′ 1" driving wheels, with 17" x 26" cylinders, which were located outside the frames, instead of the more common inside positioning, and so set a precedent that was always followed by the LTSR for its passenger engines.

They were the first engines actually owned by the LTSR, as previously all train services had been run by the Great Eastern Railway (GER) under contract. Although ostensibly designed by Thomas Whitelegg, there is some proof that the design was actually produced by William Adams, who was then actually working for the GER. Adams is famously known for inventing the 'Adams axle', a radial axle that he incorporated in his designs for the London and South Western Railway (LSWR).

Subsequent LTSR 4-4-2T engines – 37-48 – were built between 1897-8 and they had 6′ 6" driving wheels, with 18" x 26" cylinders. They were all later rebuilt with larger boilers and 19" x 26" cylinders. Engines 51-68 were built with similar dimensions, but they had bigger boilers and were built in two batches in 1900 and 1903. Engines numbered 79-82, which had 19" x 26" cylinders and 6′ 6" driving wheels, were built in 1909. The last batch had even larger boilers than their predecessors, although the rebuilt 37 class were later given the same type of boiler and were the largest tank engines on the LTSR at that time.

These engines all proved to be very successful and efficient. They were always worked hard, having to work the heavy London to Southend suburban trains, often weighing up to 400 tons, over a route abounding in speed restrictions and gradients which, although they were not exceptionally severe, were in the most inconvenient situations. The final LTSR type proved so efficient that the London, Midland and Scottish Railway (LMSR), which had absorbed the LTSR in 1912, introduced its own engines of the Atlantic wheel arrangement between 1923 and 1930 but with poor results. Many Atlantics moved far from the Essex suburban lines of their inception and some were put to work in the Midlands, with three of them going to Dundee and four of them into store at Carlisle. Interestingly the first British 4-4-2 tank engines in this country – as well as the world – were of LTSR design!

From 1898, the Great Northern Railway (GNR) went 'Atlantic mad' and began a long phase of 4-4-2 construction. H.A. Ivatt of the GNR built a total of sixty of his 4-4-2T engines, in six batches of ten each between 1898 and 1907, for use on local and commuter trains on the GNR's North London services and also for services in Yorkshire.

The GNR Atlantic tank type engine had 5′ 7$^{1}/_{2}$" coupled wheels, 175lb psi working pressure with 17$^{1}/_{2}$" x 26" cylinders. After building forty engines, Ivatt increased the cylinder diameter to 18", for the final twenty engines, with the last one being built in 1907. The first ten engines, 1009, 1010 and 1013-20 were sent to the Leeds area and the

next twenty versions, 1501-20 went to London. All but one had a short chimney and dome for underground working, as well as condensing apparatus. The exception, 1501, was fitted with a standard chimney and dome and consequently had its condensing gear removed and was sent to work at Hatfield. In spite of the advent of the more powerful 0-8-2T and 0-6-2T engines, the 4-4-2T versions remained on the King's Cross services for some years. The London and North Eastern Railway (LNER) classed these very useful engines as their C12 class and drafted them all over its system; they could be seen in places as widely separated as Hatfield, Bradford, Chester and Louth.

In May 1898, the same month and year that the fourth member of Ivatt's Atlantic tank engine design entered service, the first example of an outside-cylinder, 4-4-2, Atlantic type, express passenger tender engine was introduced to the United Kingdom. It was designed by Ivatt for service on the GNR and was numbered 990. It entered service just a few months before the first inside-cylinder 4-4-2 version designed by the Lancashire & Yorkshire Railway (LYR), was released. The Ivatt examples were

When the first ten examples of the GNR's Atlantic tank engines were completed, they were sent to the Leeds area with the next twenty examples, 1501-1520, going to London. Here we see 4511, which was one of the London batch, after it had had 3000 added to its number by the LNER at the Grouping. (Author's Collection)

appreciably larger in size than the preceding Ivatt 4-4-0s and Patrick Stirling's Singles. N° 990 had a large grate area of 26.75sq ft in a long and narrow firebox.

A further ten machines of similar design were built in 1900 and one four-cylinder 'simple' variant arrived in 1902. Ivatt then switched over to a prototype engine with a much larger boiler and a very different firebox, after which 990 and its fellows became known generally as small boiler Atlantics, with the new engines known as large boiler Atlantics. Apart from another ten small boiler engines that were built in 1903, Atlantic construction thereafter was confined to the large boiler type on the GNR. Ninety-four examples were constructed over the years 1902-10 and consisted of ninety-one standard two-cylinder 'simple' examples and three four-cylinder compound variants. The small boiler Atlantics were nicknamed 'Klondikes' after the famous gold rush which was contemporary.

The first member of the new large boiler Atlantics to be constructed was N° 251 and it was completed at Doncaster Works in December 1902. The cylinders, motion, wheel sizes and divisions were all similar to those of the small boiler Atlantic 990 class engines, but there was a slight variation to the main and auxiliary frames at the rear end. The major changes in the boiler and firebox led to dimensions which were exceptional by British standards of the day. With the length between the blast-pipe and the firebox backplate the same as the 990 series, the layout of the new boiler gave an increase of an amazing 72 per cent in evaporative heating surface. The GNR's pioneer large boiler Atlantic 4-4-2 express passenger engine N° 251, was built by Doncaster Works. The large boiler Atlantics were an unrivalled success and remained in service until the early 1950s.

Ivatt's large boiler Atlantic 4402 is seen here at speed with a long rake of mixed coaching stock. 4402 was built at Doncaster Works in June 1905 and was given the works number of 1078. Withdrawn from service on 5 August 1947, it was then scrapped at Doncaster Works. (Author's Collection)

Almost as soon as Herbert Nigel Gresley was appointed the Chief Mechanical Engineer of the GNR, he started to think about large express passenger engines. Ivatt, his predecessor, had already experimented with compound four-cylinder Atlantics with high pressure boilers and wide grate areas. Gresley's first designs followed these ideas and an old Atlantic engine was suitably modified in 1915. This was followed by plans for two different Pacific designs, one of which was merely an elongated modification of his Atlantic design, although this design was not adopted. The design that was adopted was his 1470 class Pacific 4-6-2 engines, an original design of Gresley. The pacifics became the next development of steam engines after the highly successful Atlantics.

Only three engines were ever officially named by the GNR. The first was the small boiler Atlantic N° 990, *Henry Oakley*, which was named after the GNR's General Manager and happily this engine survives in the National Collection. The other two engines were the first of Gresley's 4-6-2 Pacifics; *Great Northern* and *Sir Frederick Banbury*. The next engine to be named appeared after the formation of the

The design that was adopted as the next development of express steam engine after the highly successful large boiler express Atlantics, was the 1470 class, renamed as A1 class Pacifics, which were an original design of Gresley. Two examples were completed by the GNR, with the third one becoming the first steam engine to be completed by the newly formed L&NER. Originally numbered as 1472, the engine eventually became the world famous N° 4472, *Flying Scotsman*. Here is the 'power and the glory' of *Flying Scotsman*, seen to good effect in this image, as it prepares to depart from Newcastle Station in the mid-1930s. (W.B. Greenfield, courtesy of the NELPG)

The Lancashire and Yorkshire Railway's John Aspinall was pipped to the post at completing and releasing to traffic his version of the 4-4-2 Atlantic type express engine to the tracks, which in this case, was introduced in 1899, with the GNR getting there in 1898. The speed of the L&Y's 4-4-0s had earned them the nickname of 'Flyers', so these much larger engines were nicknamed 'Highflyers', because of their high-pitched boiler and large driving wheels. They were the mainstay of the company's premier express turns right up until the 1920's. Here is 711 during the early part of the twentieth century. (Author's Collection)

LNER in 1923 at the Grouping and this was the immortal *Flying Scotsman*! Under the LNER's classification scheme, 990, *Henry Oakley*, became a C2 class engine and was renumbered to become 3990. It was withdrawn from service from Lincoln depot in October 1937 and is now preserved by the National Railway Museum.

We now take a look at some of the other Atlantic-type engines that ran on Britain's railways.

After Ivatt's GNR 'Klondikes' of 1898 were introduced, the LYR's Class 7 introduced its class of Atlantic passenger steam engines, which were built to the design of John Aspinall. Forty examples were built between 1899 and 1902. As a result of having a high-pitched boiler that was supposed to increase stability at speed, these engines were known as 'High-Flyers'! At the Grouping of 1923, they all passed into the ownership of the LMSR and became the LMS's only Atlantic tender engine class. The LMSR gave them the power classification of 2P. Withdrawals of the class started in 1926 and the last of the class was withdrawn in 1934.

William Adams developed the 4-4-2T type of engine into his successfully famous suburban 415 class engine, which were introduced in 1882 for services on the LSWR and were originally rostered for suburban traffic. With the trailing wheels forming the basis of its 'Radial Tank' name, the class became known as 'Adams Radials'. The first twelve examples were built in 1882 by Beyer, Peacock & Co Ltd and were followed by eighteen more the following year from Robert Stephenson & Company. All these engines had $17^{1}/_{2}$" x 24" cylinders and a boiler working at 160lb pressure. The water tank capacity of 1,000 gallons was increased to 1,200 gallons in the later engines but they were otherwise similar engines. They were built in 1884-1885, with twenty built by Dub's, ten by Stephenson's and eleven by Neilson's, making a total in all of seventy-one engines. All were withdrawn between 1921 and 1928, but three examples – British Railways 30582, 30583 and 30584, which were originally the LSWR's 125, 488 and 520 – worked the Lyme Regis branch line for many years. The final example of an Adams

While the wheel arrangement and type name Atlantic came to fame in fast passenger services between railroads in the United States by mid-1895, the tank engine variant first made its appearance in the United Kingdom in 1880, when William Adams designed the 1 class 4-4-2T of the London, Tilbury and Southend Railway – LT&SR. This contemporary image shows the evolution of the tank engine on the LNWR, leading up to its version of the Atlantic tank engine. (Author's Collection)

Radial, 488, is preserved on the Bluebell Railway.

Apart from the Inner Circle service of London's underground railway, other routes circumnavigated London, although not forming a complete loop. From 1872, the London & North Western Railway (LNWR) began an Outer Circle service. Trains ran from Broad Street to Mansion House and ran via Willesden Junction and Earl's Court, diverting a service that had previously run to Victoria. The first 4-4-2 tank engines owned by the LNWR were used to work these Outer Circle trains. Needing engines suitable for this work and other London suburban traffic, they acquired sixteen of the famous Metropolitan 4-4-0 tank engines, between 1868 and 1876. In 1892, Webb converted ten of them to the 4-4-2 wheel arrangement, by removing the condensing gear, fitting cabs and greatly enlarging the coal bunkers. Eventually they found their way to various rural branches.

During the time that the LNWR were operating on the London Underground system, another LNWR 4-4-2T type engine was under construction. In 1904, George Whale, who was a British railway engineer born in Bocking, Essex, and educated in Lewisham, began building his famous Precursor 4-4-0 express engines for the LNWR. Two years later there began to appear from Crewe his version of a 4-4-2T engine, which became known as Precursor tanks. Fifty engines were built and they had 6' 3" driving wheels, 19" x 26" cylinders and a boiler pressure of 175lb. They proved to be very reliable engines and were used over most of the LNWR system and virtually took over the Euston suburban train services. They were displaced from these services during 1932, with some engines being stationed at Oxenholme just before the Second World War. The last of the class was withdrawn in March 1940.

The 4-4-2T type engine was amongst the first type to be built by the newly formed Great Central Railway (GCR) after the 1923 Grouping. The first example of eight appeared from Gorton Works in 1903, as its 9K class, later becoming the LNER's C13 class. Vulcan Foundry also built twelve examples in 1903, followed by twenty more from Gorton over the next couple of years. These engines were slightly larger than those built by the GNR, although other dimensions were similar. They had 5' 7" coupled wheels, 18" x 26" cylinders and had 160lb psi working pressure boilers. In 1907, twelve slightly modified engines were constructed and were classified by the GCR as their 9L class and later they became the LNER's C14 class. In these versions the water capacity was increased from 1,450 to 1,825 gallons and the bunker capacity was increased from 3 ton 13cwt of coal to 4 ton 6cwt, increasing the weight of the engines in working order by 44 tons. These fifty-two engines were used all over the GCR and were also used on the lines of the Cheshire Lines Committee. They were largely displaced from the Marylebone services by the Robinson A5 class 4-6-2T engines, which appeared in 1911, but when the LNER took over the steam

The last numerical member of the I2 class was N° 35, which is seen at Brighton depot with the abbreviated version of the owning companies name – having the ampersand and the R removed. By this time the brakes, which had previously been fitted to the bogie, had been removed. (Author's Collection)

engines of the Metropolitan Railway, several 4-4-2T came back to Neasden to replace the 'Met' engines sent north. They didn't remain long at Neasden and eventually returned to the Chesham branch.

Immediately before the 1923 Grouping, there was speculation in several railway publications as to whether the London, Brighton & South Coast Railway (LBSCR) eventually intended to operate all of its trains with tank engines. If this had come about, it would have been due largely to the great success of that railway's well-known I3 series of 4-4-2 tank engines designed by Douglas Earle Marsh, of which the first was turned out of Brighton Works in September, 1906 and which was classified as I1 class. Nine similar engines were built in the next nine months and a second batch of ten followed in 1907. All engines had 17" x 26" cylinders and 170lb psi pressure, but the earlier batch, 595-604, had coupled wheelbases of 8′ 9", whereas the later engines, numbered 1-10, had this cut down to 7′ 7". Between 1925 and 1932, all the engines were re-boilered and were re-classified as I1X class. During 1907 and 1908, ten similar engines known as I2 class were built, but all were withdrawn between 1933 and 1939 as they proved to be very unsatisfactory engines.

The first of the engines, N° 21, was built in 1907 and was given a 19" diameter cylinder, but this was increased to 21" in 1908, and in early 1909 this increase was also applied to engines 22-26. The later 1909 and early 1910 products, 27-30, 75 and 76, reverted to the smaller diameter cylinders, but later engines, 77-81, were given the larger diameter cylinders. The final engines of the I3 class, which appeared in 1912 and 1913, were numbered 82-91 and all had the 21" cylinder diameter.

It was a member of the I3 class, which featured in the 1909 engine exchanges between the LBSCR and the LNWR. The LBSCR's I3 class N° 23 and the LNWR's Precursor 4-4-0 N° 7, *Titan*, operated the 'Sunny South Express' service jointly between Rugby and Brighton. Anxieties about the Brighton tanks' ability to run from Rugby to Willesden without refuelling were

The final engines of the I3 class, which appeared in 1912 and 1913, were numbered 82-91 and all had 21" diameter cylinders. Southern liveried I3 class 4-4-2T Atlantic 2088 is seen resting between duties at Stewarts Lane depot. Completed at Brighton Works during November 1912, it was allocated to Brighton depot in 1947 and again in 1950, from where it was withdrawn from service and was scrapped at Brighton works in November 1950. (Author's Collection)

groundless, for the I3s were among the first British engines to be superheated and the water and coal consumption of the LBSCR engines were markedly lower than those of the Precursor. The next class of LNWR passenger engines, the George the Fifths, were fitted, like the I3s, with Schmidt superheaters. The I3s were the best and the largest of the LBSCR 4-4-2T types. Another five 4-4-2T engines of the I4 class, were built in 1908 and had 160lb psi pressure and 20" x 26" cylinders, but these were all withdrawn from service by 1940.

Like the LNWR, the Great Western Railway's (GWR) County tanks were designed on similar lines to the 4-4-0 engine. They appeared between 1903 and 1905 and had 18" x 30" cylinders, 6′ 4" coupled wheels and had a working boiler pressure of 200lb psi.They had a rather large proportion of their driving wheels exposed and proved to be very useful engines, performing good work with suburban trains around London, Birmingham and on many rural routes. They were, however, doomed by the GWR's preference for six-coupled engines and were all scrapped by 1935.

Their early success, however, was

The L&NWR's Precursor class 4-4-0 N° 7, *Titan*, is waiting at Brighton Station ready to back down to couple up with the northbound 'Sunny South Express'. The engine was taking part in C.J.B. Cooke's engine interchange trials and is seen on 22 January 1910. (L&NWR Society Collection)

The GWR's engineer, George Jackson Churchward, bought three French-built four-cylinder compound engines to allow the evaluation of the benefits of compounding against his Star and Saint class engines. The first engine, N° 102 *La France*, was delivered in 1903, with two further engines, 103 and 104, being purchased in 1905. Here is 103, President, in rebuilt form with a GWR pattern boiler in the 1920's.
(Author's Collection)

such that the GWR built one other 4-4-2T with similar features, except that the driving wheels were one foot less in diameter. This was N° 4600 and it was tried out for a time at Paddington. But it was then sent to the provinces and was cut up in 1925, without the type being multiplied. Several 4-4-2T engines were taken over by the GWR from the Welsh railways at the Grouping, six of which were from the Taff Vale Railway (TVR) and others were from the Brecon & Merthyr Railway (BMR). The former had 17" x 26" cylinders, 5′ 3" coupled wheels and 160lb boiler pressure and all were scrapped by 1926. The latter had 17½" x 24" cylinders, 5′ 6" driving wheels and 140lb psi boiler pressure and they went in 1922.

The smaller railways also had some 4-4-2T engines and for example the North Staffordshire Railway (NSR) had seven K class engines, of which numbers 8, 45, 46 and 155 were built in 1911 and 13, 14 and 39 arrived in 1912. Renumbered as 2180-6 by the LMSR, they were cut up at Crewe in 1935. They had coupled wheels of 6′ diameter, 160lb psi working boiler pressure and 20" x 26" cylinders.

The Furness Railway's 4-4-2T engines N° 38 and N° 39 were built by Kitson & Co, Leeds, in 1915 and numbers 40-43, by Vulcan Foundry in 1916, with 5′ 8" coupled wheels and 160lb psi working pressure.

Three very interesting 4-4-2 tank engines were built and owned by the Midland & Great Northern Joint Railway (MGNJR) and were constructed at Melton Constable, N° 41, in 1904, N° 20 in 1909 and N° 9 in 1910. They had 6′ coupled wheels, 160lb psi working boiler pressure and 17½" x 24" cylinders. N° 20 was scrapped in 1942 and the others in 1944. N° 41 incorporated a number of parts from some of the well-known MGNJR Beyer, Peacock 4-4-0 engines. The 4-4-2T engines were to be seen running in the Cromer area in the early thirties, painted in the attractive MGNJR ochre livery.

The 4-4-2T engine design also migrated to Scotland, but seems to have found little favour there, as only one type was built and this was a comparatively late design. It was the Reid North British Railway (NBR) class, which was built in three batches. The first thirty examples appeared in 1911 from the Yorkshire Engine Co Ltd with 18" x 26" cylinders and 175lb psi working boiler pressure and they became the LNER's C15 class. The second batch of fifteen engines appeared in 1915 with cylinder diameters increased by 1", but the working pressure was reduced by 10lb psi. Six identical engines were built in 1921. Both of the two final

When first built, the LB&SCR's H1 class Atlantics, 37 and 38, were plagued with overheating trailing axle-box problems, but once this was overcome the class as a whole seldom spent much time out of traffic and during 1906-09 they ran high mileages. Here is H1 class member 37, working a prestige Pullman service past Balham Intermediate signal box.
(Author's Collection)

batches, which were later classed as C16 class, were built by the North British Locomotive Co Ltd. They performed good work all over the old NBR system, including the Edinburgh and Glasgow suburban services, numerous country routes and the Scottish railway's lines, which penetrated over the Border.

In 1903, for use in comparative trials against his own designs, George Jackson Churchward of the GWR purchased three of the French-built Alfred de Glehn design compound 4-4-2 Atlantic engines. The first became the GWR's N° 102, La France, which was followed by two larger engines in 1905. Fourteen of Churchward's two-cylinder 2900 or Saint class engines were subsequently either built or rebuilt with this 4-4-2 wheel arrangement, including one four-cylinder, 4000 or Star class engine N° 40, *North Star*. Subsequently, all of these engines were rebuilt as 4-6-0 engines.

Wilson Worsdell of the North Eastern Railway (NER) designed his V and 4CC classes between 1903 and 1906, while John G. Robinson of the GCR introduced his 8D and 8E classes of three-cylinder compound Atlantic design engines in 1905 and 1906.

The LBSCR's H1 class version was introduced by Douglas Earle Marsh in 1905 and 1906 and was copied from the plans of the GNR's designed C1 class large boiler engines, but with minimal alterations. In 1911, L.B. Billinton was granted authority to construct a further six examples, incorporating Wilhelm Schmidt superheaters and this version

Ninety-four examples of the GNR's large boiler Atlantics were built at Doncaster Works and remained in service until the early 1950s. This atmospheric shot of King's Cross Station in September 1938, shows large boiler Atlantic 4432 reversing out of the station, with another example of the class on the left waiting for its next turn of duty.
(Antony M. Ford Collection)

became the H2 class.

William P. Reid of the NBR, built twenty examples of his I class between 1906 and 1911 and they were later known as the LNER's C10 class.

William Worsdell's successor on the NER was Vincent Raven and he introduced his V1 and Z class Atlantics between 1910 and 1917. However, by 1918 the 4-4-2 type had been largely superseded by the 4-6-0 type in the UK.

Judging by the numbers built, the British 4-4-2 type of engine had its heyday just before the First World War, when about 460 examples of the type were in service. This continued until the mid-1920s, when they began to be withdrawn and the Grouping era saw an end of new designs. The type, however, that was considered by most to have been the most successful design was surely Ivatt's large boiler Atlantics. Constructed for the GNR they really set the scene for this type of engine having been designed using ideas from the Baldwin Locomotive Co in America.

The discovery of a GNR style Atlantic boiler, which had been used at a timber drying plant in Essex, prompted the Bluebell Railway to buy it for the re-creation of the Brighton Atlantic passenger tender engine *Beachy Head*. It is hoped that in the not too distant future a total of three Atlantics will survive.

Although revered in their own right, the Atlantics were the precursors to the development by Gresley of his very successful Pacific design of engines, of which *Flying Scotsman* and *Mallard* became so very famous.

Chapter 1

WHAT IS AN ATLANTIC?

To call a steam engine an Atlantic is to use a commonly approved railway shorthand to describe the wheel arrangement of this particular type of steam engine. The system was originally devised by Frederick Methvan Whyte of America.

Whyte, who was born on 2 March 1865 and died in 1941, was a mechanical engineer who worked for the New York Central Railroad (NYCR) also known as 'The New York Central' or simply 'The Central' and was a railroad operating in the north eastern part of the United States. He is most widely known as the person who developed the 'Whyte Notation', which is used to describe the different wheel arrangements of steam engines from 1900. The notation first came into use in the early twentieth century and was encouraged by an editorial in the American Engineer and Railroad Journal of December 1900.

So this is how the system works. Frederick Whyte's system first counts the number of leading wheels, then the number of driving wheels and finally the number of trailing wheels of a steam engine, with the groups of numbers being separated by dashes. So that in the 'Whyte Notation', as it is known, an engine with two sets of leading axles – that is the four wheels in front, followed by two sets of coupled driving axles – that is four wheels and finally one trailing axle – which has two wheels, is classified as a 4-4-2.

The first use of the 4-4-2 wheel

In the early 1890s, the ACL was interested in buying an engine with more steaming capacity than their current 4-4-0s. So in 1894, Baldwin designed for the ACL an engine with a 4-4-2 wheel arrangement to carry a larger firebox and named the class after them – Atlantics. Other railroads bought this design of engine and also called their engines Atlantics. Here is N° 1027, a 4-4-2 Atlantic of 1896, built by Baldwin Locomotive Works for the Atlantic City Line. This type were also known as a Camelback because the cab sat on the boiler. This particular engine was fitted with Walschaerts valve gear and was also superheated. (Author's Collection)

At the end of the nineteenth century many American railroads preferred the 4-4-2 Atlantic wheel arrangement, due to its ability to provide fast and reliable services. Here is an image of a Baldwin Locomotive Works 4-4-2 Camelback design, which was taken in 1907 and shows the Philadelphia Railroads N° 342. The driver (or engineer in American parlance), occupied the cab straddling the boiler, with the fireman being located between the firebox and the tender loading the firebox. The Wootten firebox was made very wide to allow the combustion of anthracite coal waste, known as Culm; indeed it was so wide that it was provided with two firing openings. (Author's Collection)

arrangement for a tender engine was under an experimental double-firebox engine, which was built to the design of George Strong at the Hinkley Locomotive Works in Boston, Massachusetts, in 1888. But the engine was not successful and was scrapped soon afterwards.

The Atlantic wheel arrangement was named after the second American tender engine class, the first being the 4-4-0, built by the Baldwin Locomotive Works in 1894, for use on the 'Atlantic City Line' of the Philadelphia & Reading Railroad (PRR).

The Baldwin Locomotive Works ideas on 4-4-2 tender engine were soon copied in the United Kingdom, initially by H.A. Ivatt of the GNR with his small boiler Atlantic of 1898, which became known as 'Klondikes', after the Yukon gold rush of the same era. These were quickly followed by the LYR Class 7 Atlantic passenger steam engines built to the design of John Aspinall. Forty were built between 1899 and 1902 and they were known as 'High-Flyers' as a result of having a high-pitched boiler that was supposed to increase stability at speed.

On 20 July 1904, train N° 25, a regularly scheduled train that ran from Kaighn's Point in Camden NJ to Atlantic City NJ, with PRR P-4c class 4-4-2 N° 334 and five passenger carriages, set a new

In the early 1890s, many American railroads adopted the 4-4-2 Atlantic wheel arrangement, as these engines provided a fast and reliable service to such places as Atlantic City and New Jersey. Here is N° 350, a 4-4-2 Atlantic built for the Pennsylvania-Reading Seashore Lines. The photo shows the 4-4-2 Atlantic wheel arrangement and the wide Wootten style firebox, which gave inspiration to the GNR Atlantics. (Author's Collection)

This small boiler Atlantic N° 3949 was constructed at Doncaster Works in March 1900 and was given works number 872. It had a superheater fitted in August 1917 and was fitted with piston valves in April 1923.
(Author's Collection)

speed record. It ran the 55.5 miles in orty-three minutes at an average speed of 77.4mph. The 29.3 miles between Winslow Junction and Meadows Tower outside of Atlantic City, were covered in twenty minutes at a speed of 87.9mph. During the short segment between Egg Harbor and Brigantine Junction the train was reported to have reached a speed of 115mph! This claim of course was based on the visual reading of a speed gauge and there was no 'paper trail' to support this claim.

The development of Atlantics was continued with those built by Ivatt with his small boiler and large boiler Atlantic engines; tank versions were also built. The Atlantic wheel arrangement was also adopted by Douglas Earle Marsh (1862-1933) when he moved

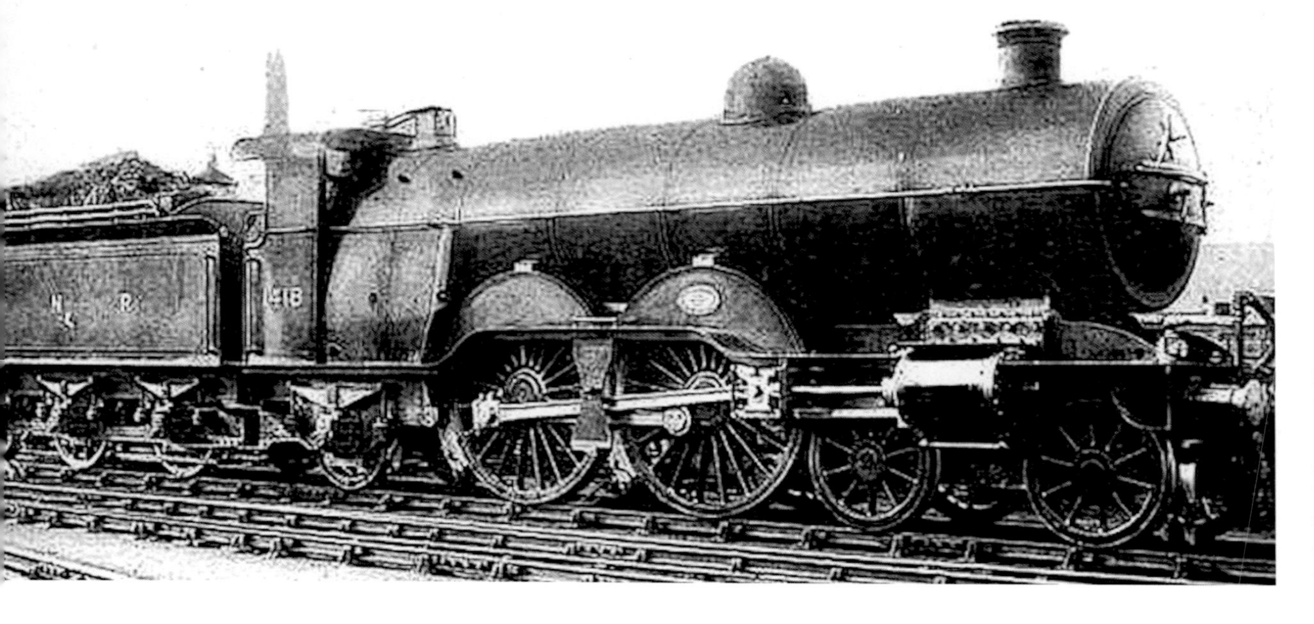

Large boiler Atlantic N° 1418 was built at Doncaster Works and was completed in May 1906. This photo clearly shows the 4-4-2 Atlantic wheel arrangement and the wide firebox, which gave the GNR Atlantics such extra power and speed compared to contemporary engines.
(Author's Collection)

H.A. Ivatt of the GNR built a total of sixty Atlantic C12 class 4-4-2T engines in six batches of ten each, between 1898 and 1907, for use on local and commuter trains in Yorkshire and North London. They were built at Doncaster and were originally designated by the GNR as their C2 class. Seen here is N° 4532, which was completed in December 1903 and which lasted until April 1937. (Author's Collection)

The Midland and Great Northern Joint Railway (MGNR) owes its name to the two older companies which owned it before they became merged into the LMS and the LNER respectively. The MGNR too used the Atlantic tank wheel arrangement and in 1904 its Melton Constable Works undertook the production of its own 4-4-2 tank engines. They had two outside cylinders of 17$\frac{1}{4}$" x 24", 3' 0" diameter bogie wheels, 6' 0" diameter driving wheels and 3' 6$\frac{1}{2}$" trailing wheels. The boiler had a round-top firebox, with a total heating surface of 1,099sq ft. Boiler pressure was at 160lb psi and the total weight in working order was 68 tons 9cwt. The chimney was of the Melton third pattern design. The tank capacity was of 1,650 gallons of water and had a 2 ton coal capacity. Three 4-4-2T engines were built in 1904, 1909 and 1910 and saw service until 1944. After these 4-4-2T engines were built, no further engines were built at Melton, but a great many modifications were always being made. N° 41 was built in 1904, N° 20 in 1909, and in 1910 the third one, N° 9 – seen in the image. They were used between Lowestoft and Yarmouth Beach shed, the Potter Heigham local service and the Cromer Beach to Melton Constable trains. The tanks originally had a square profile, but sloping fronts were fitted between 1933 and 1934 to improve the view from the cab. (Author's Collection)

Nº 598, was part of the batch of the first ten I1 class 4-4-2 tank Atlantics that were designed by D.E. Marsh for the LBSCR. The batch were numbered 595-604 and incorporated the wheels, coupling-rods and parts of the motion from older D1 and D2 class engines, which had been taken out of service. (Author's Collection)

This view shows large boiler Atlantic Nº 4426 at Sheffield, with the actual arrival of the first Sheffield Pullman service, on 2 June 1924. (Antony M. Ford Collection)

from the GNR to the LBSCR, where he became the Locomotive, Carriage & Wagon Superintendent, from November 1904 until his early retirement on health grounds in July 1911. A total of eleven Atlantics were constructed for the high speed luxury services of the Brighton line, where they were much admired and appreciated.

The name Atlantic became a magic word in the world of railways as it conjured up speed and power and all railways wanted to have Atlantics in their collection, so that they could also ride on the band wagon of this famed class of steam engines.

Large boiler Atlantic N° 1438 was completed in March 1908. It was renumbered by the LNER in 1923 and became N° 4438, as is seen here. When it was renumbered a second time by the LNER, it then received the number N° 2868. (Author's Collection)

Large boiler Atlantic N° 3299 was built in May 1905. It never carried its second LNER number and in June 1930 it was allocated to King's Cross depot. (Author's Collection)

Chapter 2

WHY WERE THE GREAT NORTHERN ATLANTICS BUILT?

By the 1840s, it was possible to travel from Yorkshire to London by rail, but the journey was a long and circuitous route travelling via Derby and Rugby to Euston. The LYR proposed that a more direct route should be constructed from London, heading directly north and passing through Peterborough, where there would be a loop line to Boston, Lincoln and Grantham, and at Doncaster there would be branches to Sheffield and Wakefield. Joseph Locke, the renowned surveyor and civil engineer, performed the initial survey, although his proposal did meet with some very strong opposition from a group led by George Hudson, who was notorious as being the so-called 'Railway King'. George Hudson employed every kind of delaying tactic that he could think of, so much so that Joseph Locke was forced to resign his commission at very short notice. In a few days he was replaced by another renowned civil engineer, William Cubitt. As soon as the bill was passed by parliament, the 'London & York' name was dropped and the new title of the GNR was adopted in its place.

Plans for the terminal station at King's Cross were first made in December 1848, under the direction of George Turnbull, who was also the resident engineer for the construction of the first 20 miles of the GNR route out of London. After the detailed design was completed by Lewis Cubitt, the construction was then carried out between 1851-2, on the site of a former smallpox hospital. The main part of the station, which today includes Platforms 1 to 8, was opened on 14 October 1852 and replaced the temporary terminus which had

George Hudson, 'The Railway King', was one of the great financiers of the nineteenth century. His story is linked with that of the GNR, where he spent £3,000 a day in legal fees fighting rival companies. (Author's Collection)

been built at Maiden Lane and had opened on 7 August 1850.

The area of King's Cross where the station was built was previously a village known as Battle Bridge, based around an ancient crossing of the River Fleet. The original name of the bridge was Broad Ford Bridge, with the name 'Battle Bridge' arising due to the tradition that this was the site of a major battle between the Romans and the British Iceni tribe, famously led by Boudicca, Britain's warrior queen. There is a belief that she is buried between Platform 9 and Platform 10 at King's Cross Station as, according to folklore, King's Cross is the site of 'Boudicca's final battle'. There are passages under the station that her ghost is reputed to haunt.

But how times change! History appears to have been rewritten, as very strangely, Platform 9³/₄ is now apparently very important to young wizards, as that is where one departs for Hogwarts Castle via the Hogwarts Express. In response to the worldwide popularity of *Harry*

King's Cross railway station opened in 1852 and was built as the London hub of the GNR and the London Terminus of the East Coast Main Line. It took its name from the King's Cross area of London, which was named after a monument to King George IV that was demolished in 1845. (Author's Collection)

King's Cross Station, London, was designed by Lewis Cubitt and was built by builders John and William Jay. The ribs supporting the roof covering were originally constructed of laminated timber, but were subsequently replaced with steel. The roof spans are 105ft wide by 800ft long. Despite the building's grand façade, the location was hugely criticised as being a '... a howling waste' and even the GNR's original Goods Manager, J. Medcalf, described the area as '... a long battalion of rag sorters and cinder beaters'. Some have said that this sentiment turned into a kind of inferiority complex, which was not helped by the construction of the ornate Gothic styled St Pancras Station next door. (Authors Collection)

The very first two Sharp Singles engines built for the GNR, N° 1 and N° 2, were delivered directly to the East Lincolnshire Railway by March 1848. They were very 'able and capable' engines when asked to handle the light and slow traffic on the level Lincolnshire lines. Here we see a very primitive Sharp 2-2-2T engine N° 1, as running in 1864, with a 10' 6" rear wheelbase. Archibold Sturrock, having progressed from being Daniel Gooch's assistant on the GWR from 1840, was the Locomotive Superintendent of the GNR from 1850 until 1866. He designed and fitted compensating levers on these primitive engines, which greatly assisted them to run on the relatively poor and lightweight track. (Author's Collection)

The greatest claim to fame of the Caledonian Railway 4-2-2 N° 123, which was built at Springburn Works, Glasgow, in 1886, was its participation in the celebrated 'Race to the North', which took place between East Coast & West Coast railway companies in 1888. After a working life of almost fifty years, 123, was stored away until the late 1950s, when BR staff decided to restore it to its former glory. In September 2010 it was moved to the new Riverside Museum in Glasgow for display. Former Caledonian Railway 4-2-2, N° 123 – which was built by Dugald Drummond in 1886 – is seen in September/October 1953, at Battersea Wharf being prepared for display for an exhibition of Royal Rolling Stock being held there. (Antony M. Ford Collection)

Potter, King's Cross has had installed an actual $9^3/_4$ sign and a 'half-trolley' lodged into a wall, so that fans could visit the station and take fun pictures!

So with the paperwork in place and the line's terminus being constructed at King's Cross, the line north to Peterborough was actually constructed by Thomas Brassey. The initial climb out of London was stiff for the period, with gradients of around 1 in 110 for some miles, but after that there were very few gradients exceeding 1 in 200 for any meaningful distance, at least on the main line.

Technological progress moved on in the design of steam engines, leading to a notable engine on express duties, that of Patrick Stirling's 4-2-2 Single class, which had 8' driving wheels. Patrick Stirling moved to the GNR in 1866 and designed his original Single in 1870. The Single engines were built for speed, power and to handle the gradients on the main GNR line from London to York. They were also built to compete against the Midland Railway and the London & North Western Railway (LNWR) in the 'Races to the North'. The races were carried out over two summers in the late nineteenth century, when British passenger trains belonging to different railway companies literally raced each other from

A period illustration of the original engine shed at King's Cross, of the GNR. (Author's Collection)

Technological progress moved the design of steam locomotives on, leading to a notable locomotive on express duties that was designed by Patrick Stirling. His 4-2-2 Single class, was a locomotive with 8' driving wheels designed in 1870. Patrick Stirling built his Single engines for speed and power and to handle the continuous gradients on the main GNR line from London to York. Here is Stirling's 4-2-2 Single N° 644, which was built in 1877 and which had been built with a much more stable design, as compared with previous examples. (Author's Collection)

London to Scotland over the two principal rail trunk routes to Scotland, the East Coast and the West Coast.

East Coast Singles engines worked the ever-increasing train loads from 1880, due to improved rolling stock and rail travel becoming more popular. So, along with the considerable improvements in technology, the overall time from London to Edinburgh was reduced to $8^1/_2$ hours by 1888, an improvement of two hours.

From 1900, the King's Cross to Edinburgh services were improved again, with the actual trains being made more customer-friendly. By 1890, demand for better facilities was met by more sleeping and dining cars, thereby introducing heavy eight and twelve wheel bogie corridor vehicles, plus trains formed of more actual vehicles and so extra weight. New features incorporated in the trains that later became standard were corridor connections between the carriages and the unheard of luxury of heating inside the passenger compartments. The introduction of dining cars included in the GNR trains was great news for passengers, as they could now take unhurried luncheon on the train if they wanted to. The bad news was that even though the York stop had been reduced to fifteen minutes, there was even more of a feeding frenzy for passengers choosing to take lunch at York, as they tried to

A contemporary view of 'Third Class comfort to Scotland', as provided by the Great Northern Railway in 1894. (Author's Collection)

A dining car as used on Anglo-Scottish express passenger services of the 1890s. The off-centre aisle gave a 'four plus two' seating table arrangement that was quite common for the time. (Author's Collection)

consume their scalding hot soup in the limited time allowed. So to add insult to injury, even though the engines made their trains move quicker and even accounting for a shorter stop, the through end-to-end journey time still stayed at 8½ hours.

Due to a long-standing agreement between the competing East and West Coast main line railway companies since the famous 'Railway Races' of 1888 and 1895, the high speeds attained on the so-called 'Scotch Expresses' were limited and the travelling time for the 392 miles between the capitals had been reduced to only a relatively pedestrian 8¼ hours, but train weights were still increasing. Stirling's Single wheeled express engines of modest boiler power, although they performed brilliant work, were not designed to handle loads soon to exceed 200 tons exclusive of engine and tender; nor were they able to meet the tighter schedules. Until the end of the nineteenth century, trains such as the *Flying Scotsman* or the 'Special Scotch Express' as it was more commonly known, took 10½ hours to complete the journey from King's Cross Station in London, to Edinburgh Waverley and the train included six-wheel coaches in its formation.

In January 1894, the GNR Board called for new engines to be built of increased power, to deal with the ever heavier loads and tighter schedules, only to be dissatisfied when no really suitable design was forthcoming. Meanwhile, the 'running department' had already resorted to costly 'pilot-running' – the process of putting a second engine in front of the principal engine and so 'double-heading' major services. However, this was costly to the railway because of the extra engines being used, together with the extra crewing, extra coal, extra water and maintenance costs. The cumulative effect of this led to the decision by the GNR's Board of Directors to replace Patrick Stirling, who was by then 75 years old, as he was not forthcoming with the desired 'magic' engine.

But the Board's prayers were answered with the appointment of H.A. Ivatt, from the following March, approved on 1 November 1895. The very same day, they asked Stirling to comment in writing on the general state of the engine stock position. However, after Stirling's sudden death on 11 November 1895, the report was made instead by the Locomotive Accountant, J.W. Matthewman. He concluded that there was an overall shortage of engine power, 51.75 per cent of the engines ran three or more double turns daily and some worked continuously for 140 hours. The mean average was 19,476 miles per engine per annum and to work at the same rate as other larger companies, the GNR would require an additional 148 engines. During the four month interregnum that followed, engines of various Stirling designs were ordered to be built, just to ease the overall situation, although, of course, it was not possible to provide new designs for the top-link express duties.

Consequently, Ivatt took office at Doncaster in March 1896 and was confronted with the enormous task of increasing the entire engine stock, not only in numbers but in power. As related in the Ivatt Policy, a gradual solution was sought in meeting problems of goods, mineral and local passenger services, though the prompt provision of domed boilers which, having a larger heating surface, improved the performance of the 8-footers and soon became an interim measure.

After Ivatt had been in office for four months, a table was prepared showing the rate at which loads had increased from 180 tons and his attention was drawn to the fact that the passenger engine field included a fleet of top-link Singles unable to tackle important East Coast Joint Stock loads unassisted, as in the previous sixteen years they had been double-headed and they were timed to run at higher speeds. However a similar situation also existed on other main line railways at that time. Most railway engine engineers still held the theory that a low-pitched boiler, having a low centre of gravity, was essential for steady running and it was not until a practical demonstration was carried out on the Caledonian Railway in 1896, that this assumption was proved to be erroneous. But a 'then-current order', for standard 2-4-0 engines gave an immediate opportunity to develop the design into a 4-4-0 type by the end of the year for rapid multiplication, to satisfy the widespread demand for secondary passenger engine duties. So a standard 4' 5" boiler was introduced for these and it was also used at a later date in a deferred scheme, to develop the Stirling 2-2-2 into a new and useful inside-cylinder 4-2-2 type.

Meanwhile, Ivatt still showed that he had faith in the 8-footers, by stating that they would be able to cope with the heaviest trains if a still larger boiler were provided. He concluded, however, that this was not feasible without a reduction in driving wheel diameter, as the space available for the boiler was dependent upon the axle height (smaller wheels = small axle height = more space for a larger boiler). It became imperative to design a different version of steam engine with a revised wheel formation to solve the heavy-loadings problem.

Seeking greater power and adhesion for his engines, Ivatt took inspiration from the American route of railway development and so with ideas from the Baldwin Locomotive Works design for the 'Atlantic Coast Line' in North America, he designed the first 4-4-2 or Atlantic type of steam engine to enter service in Great Britain. The very first 4-4-2 tender engine in the world had actually been an experimental engine and was built in 1888 by the Hinckley Locomotive Works, Boston, Massachusetts, USA, and was named A.G. Darwin. Retaining a bogie and outside cylinders, it was decided to provide four coupled wheels with reduced driving-wheel diameter, sufficient to enable a 4' 8" boiler to be placed over the driving wheels. Early in 1897, Ivatt outlined this proposal for a new elongated design, estimated to have an axle load of 16 tons and a weight of 52 tons in working order. The relevant approval was obtained to build one engine.

At the close of 1897, detail design on this unprecedented 4-4-2 type of engine was still being worked out, but the building was finally completed at Doncaster Works in May 1898, when it emerged as the first British-built Atlantic-type tender engine. Numbered 990, it was officially designated simply as the 990 class, although two outside events brought a permanent nomenclature to the engine during the course of its construction. First, contemporary American 4-4-2s had won fame for their high speed runs between Camden and Atlantic City and subsequently all engines with a 4-4-2 wheel arrangement became known as Atlantics after the famed high-speed runs. Second, the 1896 Gold Rush to the Yukon remained a worthy news item at the time and earned it the nickname of 'Klondike', after the valley of the same name. It is held in railway folk-lore that while building N° 990, a fitter carrying material across his shoulder towards the engine, was asked where he was going and jestingly replied 'Off to the Klondike'. This nickname stayed with the class until the end of its days.

N° 990 was subjected to a period of exhaustive 'road-trials' before a decision was taken to multiply the basic type and in 1898, some twenty-eight years after Patrick Stirling's first 4-2-2 had appeared from Doncaster Works, Britain's first 4-2-2 Atlantic, locomotive 990, was ready for work. There could hardly have been a more complete reversal of a predecessor's policy than the

***Henry Oakley* is** seen at King's Cross Station, with one of the gas holders that made Gasworks' Tunnel famous. (Peter N. Townend Collection)

proportions embodied in this new engine. Instead of 19½" by 28" cylinders, there were cylinders of 18½" diameter by 24" stroke and the new 4-4-2s had a heating surface of 1,442sq ft with a fire-gate area of 26.8sq ft compared previously with that of 1,032sq ft heating surface and a fire-gate area of 20sq ft on the Singles.

This resulted in Ivatt's Atlantic having a nominal tractive effort of 15,860lb, whereas that of Patrick Stirling's last 4-2-2 was 16,100lb. But the tractive force formula means nothing unless there is sufficient steam to make it effective. Ivatt had publicly proclaimed that it was his opinion that the capacity of a steam engine was 'its ability to boil water' and here was the principle in action. There was no question, of course, as to which was actually the more powerful of the two types. In total, twenty-two examples were built between 1898 and 1903 at Doncaster Works. They had a large capacity boiler and it was this extra steam-raising capacity that gave the Atlantics the edge over Stirling's single-wheelers.

The GNR's first Atlantic entered service in May 1898, with the first production Atlantics entering service in March 1900. They all proved to be fast and lively runners, indeed, Ivatt had to caution his drivers 'to rein in the speed' because stretches of the track between London and Doncaster were considered to be too uneven for safety's sake, with regards to high speed running! In their turn, the engine men would have told Ivatt that the cylinders were no match for the boiler. These first Atlantics had to be worked at undesirable and uneconomic rates to achieve the expected performance – in other words they were 'thrashed'!

In July 1902, a four cylinder version, N° 271, was completed and in December of the same year N°

Small boiler Atlantic N° 271, was the first four-cylinder engine to be owned by the GNR. It is seen here in its rebuilt form. (Author's Collection)

251 appeared as the forerunner of a new class of Atlantics. With most things from the running plate down, including pistons wheels and running gear, a development of the 'Klondike' was instigated by the provision of a much larger boiler and wide-firebox. A final batch of ten 'Klondikes' was brought out in 1903 and these together with N° 271, totalled twenty-two engines.

These new 4-4-2 engines were a great advance on the latest Stirling productions. Compared with the 1003 series '8-footers', they had larger boilers with improved heating surface and grate area through their cylinder volume was one-fifth less.

In June 1900, the British pioneer Atlantic 990 was named *Henry Oakley* in honour of the former General Manager of the GNR, who had retired two years earlier. The name was, however, applied somewhat sketchily. Separate plates bearing one word only, were attached to each wheel splasher, but were of insignificant size and were without any pretension to style or character. *Henry Oakley* remained the only named engine on the GNR until the first of H.N. Gresley's Pacifics, N° 1470, *Great Northern*, appeared in 1922.

As originally built, N° 271 was the first four-cylinder engine to be owned by the GNR, although the cylinders were small, with a volume barely 4 per cent greater than that of the two cylinders of N° 990, and N° 271 was not a success in its original form. It still remained unique after it was rebuilt in 1911, when it was fitted with two inside cylinders, instead of the more conventional outside cylinders.

In 1909, Atlantic N° 988 became the first GNR engine to be superheated. The superheater was of the Schmidt type and at later dates other 'Klondikes' received either Robinson or Gresley twin-tube superheater equipment, with the Robinson type being finally adopted for the class. The 'Klondikes' were eventually eclipsed in performance by the large-boiler Atlantics, which proved to be even more powerful engines, capable of sustained efforts within the limit of their boiler capacity. All 'Klondikes' entered LNER stock at the grouping in 1923 and were withdrawn between 1935 and 1945. *Henry Oakley* was withdrawn from service in 1937 and was repainted in GNR style for exhibition in the old LNER Railway Museum at York.

A view looking into the cab of a GNR Atlantic express passenger engine. This type of engine was the predecessor of Gresley's Pacifics. (Author's Collection)

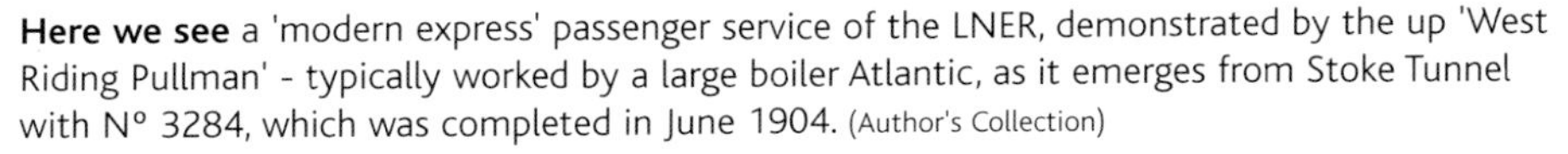

Here we see a 'modern express' passenger service of the LNER, demonstrated by the up 'West Riding Pullman' - typically worked by a large boiler Atlantic, as it emerges from Stoke Tunnel with N° 3284, which was completed in June 1904. (Author's Collection)

The GNR's Atlantic express passenger engine design was a very important link in the design and evolution of high-speed rail travel in this country and indeed the world. With the introduction of large boiler Atlantics, East Coast train services were improved even more. Here is part of the GNR's timetable of 1911, showing a large boiler Atlantic. (Author's Collection)

N° 3279 was a large boiler Atlantic built in June 1904 and was originally renumbered by the LNER at the Grouping in 1923, a secomd time, to become N° 2808. It is seen here working a luxurious Pullman service. (Author's Collection)

With the introduction of large-boiler Atlantics, East Coast train services were improved even more. The GNR's Atlantic express passenger engine was a very important link in the design and evolution of high-speed rail travel in this country and indeed the world. Gresley was also inspired by the American design of a wide

The engines were originally known in official GNR documents as the 990 class, however, upon the appearance of the large-boiler Atlantics from 1902, both types were included together in the C1 class. The load class was Y, which became Z, when superheated. The LNER then segregated them into two classes: C1 class for the large boilers; and C2 class for the 'Klondikes', including the existing inside cylinder version. From 1924 onwards, 3000 was added to their running numbers. Under the Thompson renumbering scheme formulated in 1943, the three remaining 'Klondike' survivors were allocated the numbers 2892/2893/2894, but all were withdrawn before the scheme was implemented.

When originally built in August 1908, large boiler Atlantic N° 1451 was built without a superheater, receiving one in March 1922. It was then renumbered by the LNER, first becoming 4451 and later 2881. The image shows it working the 'West Riding Pullman'. (Author's Collection)

A fascinating view of a bygone era, with a large boiler Atlantic taking centre stage while working a passenger service.
(Author's Collection)

firebox and steam raising boiler and went on to build his Pacific design A1 class Pacific design engine, leading on to *Flying Scotsman* and *Mallard* – and we all know what happened there; *Flying Scotsman* became the first steam engine in the world to reach an authenticated speed of 100mph and *Mallard* became the holder of the world speed record for steam engines at 126mph!

Ivatt's large boiler Atlantic N° 4407 is seen working a goods service through Potters Bar in 1927.
(Author's Collection)

Chapter 3

HENRY ALFRED IVATT AND THE GREAT SOUTHERN AND WESTERN RAILWAY OF IRELAND

In 1895 concern arose within the GNR management over Stirling's unwillingness to retire, despite his advancing years. Matters were further complicated by the death of John Shotton, the Locomotive Works Manager at Doncaster since October 1866 and who had been recognised as Stirling's assistant. When Stirling notified the GNR's Board of Shotton's death in May 1895, he suggested that the Board might 'consider it an advantageous opportunity at any time'. So moves were initiated towards finding a suitable candidate to replace Shotton and in due course, to take over Patrick Stirling's job.

As there was nobody in the GNR's Doncaster organisation who was considered to be at least adequate for the post, the directors had to spread their nets wider and so from some undisclosed source, they decided to consider H.A. Ivatt, who was at that time working for the Great Southern & Western Railway of Ireland (GSWRI). Indeed, the GSWRI proved to be a very good training ground for many engineers, who later migrated to the railways of Britain and proved their worth on the GNR & NER railway companies as well as the LYR and indeed the Southern Railway! Here is a very brief resume of the Locomotive Superintendents there between 1845 and 1924.

Henry Alfred Ivatt (16 September 1851-25 October 1923) was born in Wentworth, Cambridgeshire, and was educated at Liverpool College. He became the Chief Mechanical Engineer of the GNR from 1896 to 1911. Here is an autographed image of Ivatt, taken in 1910. (Author's Collection)

Locomotive Engineers of the Great Southern and Western Railway

Alexander McDonnell was an Irish locomotive engineer and a civil engineer. Born in Dublin 18 December 1829, he died in Holyhead 14 December 1904. He brought order and standardisation to workshops and produced engine designs of the Great Southern and Western Railway of Ireland. He was later employed to do the same for the NER in England. Between 1864-1883, he designed the following engine types:

GSR Class 2 or Class D19
GSR Class 21 or Class G4
GSR Class 47 or Class E3
GSR Class 90 or Class J30
GSR Class 91 or Class J29
GSR Class 92 or Class H2
GSR Class 101 or Class J15
GSR Class 203 or Class H1
GSR Class 204 or Class J12
GSR Class Sprite or Classes L4 and L5

Sir John Audley Frederick Aspinall who was born 25 August 1851 and who died 19 January 1937, was a British mechanical engineer who served as the Locomotive Superintendent of the GSWRI and the LYR. He introduced the vacuum brake system to his locomotives in Ireland, a trend which was to follow in Britain, and he designed several locomotives. He was also President of the Institution of Mechanical Engineers and of the Institution of Civil Engineers. Between 1883-1886, he designed the following engine types:

GSR Class 52 or Class D17
GSR Class 60 or Class D14

R. Coey worked between 1896-1911, but sadly very little else is known about him other than that he designed the following engine types:

GSR Class 27 or Class C4
GSR Class 211 or Class J3
GSR Class 213 or Class I1
GSR Class 301 or Class D11
GSR Class 305 or Class D12
GSR Class 309 or Classes D3 and D10
GSR Class 321 or Classes D2, D3, and D4
GSR Class 333 or Classes D2, D3, D4, and D4a
GSR Class 341 or Class D1
GSR Class 351 or Class J9
GSR Class 355 or Class K3
GSR Class 362 or Class B3 – 'Long Toms'
GSR Class 368 or Class K4

Richard Edward Lloyd Maunsell:- was born 26 May 1868, at Raheny, County Dublin, in Ireland. After graduating, he began an apprenticeship at the Inchicore works of the Locomotive Engineers of the GSWRI, under Ivatt in 1886, completing his training at Horwich Works on the LYR, just as Nigel Gresley had done before him. At Horwich, he worked in the drawing office before occupying the post of locomotive foreman in charge of the Blackpool and Fleetwood District. He then went to India in 1894, as Assistant Locomotive Superintendent of the East Indian Railway Company (EIRC), being subsequently District Locomotive Superintendent of the Asansol District. He returned in 1896 to become Works Manager at Inchicore on the GSWRI, moving up to become Locomotive Superintendent in 1911. In 1913, he was selected to succeed Harry Wainwright as Chief Mechanical Engineer of the South Eastern and Chatham Railway (SECR) and held the post from 1913, until the 1923 Grouping and then the post of CME of the Southern Railway until 1937. Between 1911-13, he designed the following engine types:

GSR Class 257 or Class J4
GSR Class Sambo or Class L2

Henry Alfred Ivatt:- designed the following engine types between 1886-1896:

GSR Class 33 or Class F6
GSR Class 37 or Class C7
GSR Class 201 or Class J11
GSR Class Jumbo or Class J13

He was born in Wentworth, Cambridgeshire, on 16 September 1851 and was educated at Liverpool College. From here, aged seventeen, Henry was apprenticed to John Ramsbottom at Crewe Works of the LNWR. Following the completion of his mechanical training, he worked as a fireman for six months at Crew North shed and was made Assistant Foreman at Stafford. From he progressed as Foreman of Holyhead locomotive depot in 1874, before being promoted to the head of the Chester District, leaving here in October 1877, to become Southern District Superintendent of the GSWRI, at Inchicore. In 1882, he was appointed to the post of Assistant Locomotive Engineer there, where he patented a design for a sprung flap for vertically-opening carriage windows that became ubiquitous.

In 1886, Ivatt succeeded J.F. Aspinall as Locomotive Engineer, when the latter left for a similar position with the LYR.

Late in August 1895, the GNR General Manager and one Director made a special trip to Dublin to visit Ivatt. Samuel Waite Johnson , Chief Mechanical Engineer of the Midland Railway, John Aspinall, Locomotive Superintendent of the LYR, Francis William Webb, Chief Mechanical Engineer of the LNWR and William Dean the Chief Locomotive Engineer for the GWR provided references for him. These proving favourable, Ivatt was asked to meet the GNR's Chairman in London early in October, when he was offered the Doncaster vacancy. At a Board meeting held at King's Cross on the 1 November 1895, approval of his appointment as the next Locomotive Engineer was made in the presence of Patrick Stirling, who immediately upon returning to Doncaster after the meeting, took to his bed and ten days later, died

Ivatt took up his new appointment at Doncaster in March 1896. The immediate problem was to build up the GNR's engine stock to such a point, that the larger engines were able to handle increased loads whenever traffic demanded it. Patrick Stirling had left the locomotive department in a

A fine sight of one of Ivatt's express Atlantic engines N° 3285. When it was withdrawn from normal revenue working service on 2 May 1946, it was put to further use by carrying out Stationary Boiler duties at Doncaster Works, from September 1947, until March 1952. (Author's Collection)

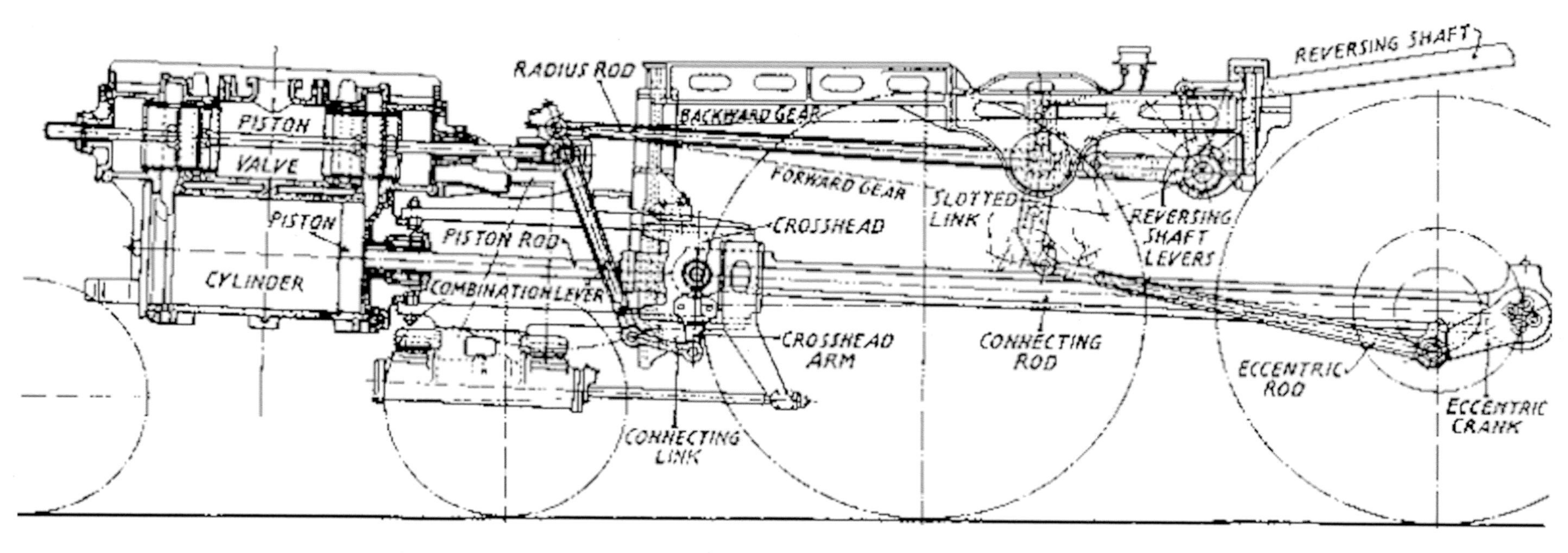

Walschaerts valve gear uses eccentrics, which convert rotary into reciprocating motion. The Caprotti and Lentz type systems utilise poppet valves – also called mushroom valves, together with camshafts. In the case of H.N. Gresley's conjugated valve gear, which was used on his three-cylinder engines, the valve of the inner cylinder derives its motion from extensions on the outside valve spindles. O.V.S. Bulleid used a chain-drive system, which is immersed in an oil-bath for the valve gear on his light Pacific engines. (Peter N. Townend Collection)

Here we see H.A. Ivatt's large boiler Atlantic locomotive, No 4421, in LNER days. Completed at Doncaster Works on 31 August 1907, it was given works 1166 and was interestingly withdrawn exactly thirty-seven years later on 31 August 1944. (Author's Collection)

well ordered and economical state; however, his small engines were rapidly becoming inadequate. During his fifteen years at Doncaster, Ivatt succeeded in the task of providing larger and more powerful engines, which to a large extent incorporated the use of standard interchangeable components between the different classes.

At the GNR, he became associated with the 4-4-2 Atlantic type engine, which he introduced to Britain from North America and his large-boiler version of these engines proved remarkably adaptable to the demands made of them over their years in service.

Ivatt was also the first engineer to introduce Walschaert's valve gear into Britain. Planning to retire at the age of sixty, he tendered his resignation in the autumn of 1911 and retired on 2 December 1911. He was succeeded as Chief Mechanical Engineer of the GNR by Herbert Nigel Gresley.

Henry Ivatt had six children, the first of whom, Campbell, died as a child in 1898. His son, George, was also a locomotive engineer and was the post-war CME of the LMSR. His daughter, Marjorie, married Oliver Bulleid, the Chief Mechanical Engineer of the Southern Railway. Ivatt died in Haywards Heath, Sussex 25 October 1923 aged seventy-three.

When it was originally built in March 1906, C1 class 4-4-2 N° 1414, received slide valves and kept them until it was withdrawn, as N° 4414, in October 1944. (Author's Collection)

In immaculate condition, C1 class 4-4-2 large boiler Atlantic N° 4405 is seen, along with an A1 class 4-6-2 Pacific and a member of the C12 class 4-4-2T, ready for display at an open day. N° 4405 was another of the few which succumbed to the cutter's torch at Darlington North Road Works. (Author's Collection)

A fine action shot as large boiler Atlantic N° 3284 'takes-up' water and drenches the first two Pullman Cars in this LNER era picture. (Author's Collection)

Chapter 4

THE AMERICAN CONNECTION

The first steam locomotive proper which came into use was Richard Trevithick's engine of 1804, the boiler of which had a cylindrical cast iron shell with a wrought iron internal cylindrical return flue. Obviously, no large amount of heating surface could be obtained with any such design and so few improvements were made in the following years until the multi-tubular system was introduced.

The multi-tubular system was adopted by George Stephenson in the boiler of his 1829 engine, *Rocket*. The multi-tubular boiler is the basis for all subsequent modern steam locomotives, outside of the firebox design. In 1826, a Mr Neville took out the English patent for a vertical tubular boiler, which stated that the system was equally applicable to horizontal boilers. It would seem, therefore, that this portion of the design of steam locomotives is owed to the vertical land boiler. However, the multi-tubular system has also been credited to a Mr Henry Booth, who was the treasurer of the Liverpool & Manchester Railway (LMR) and it is believed that he suggested to Robert Stephenson that a multi-tube boiler should be used.

George Stephenson improved his original boiler design for *Rocket* in 1833, by projecting the cylindrical portion of the shell over the firebox, which of course allowed a much larger steam space, as well as a much greater steam raising surface. This modification of Stephenson's original design was the basis on which just about all subsequent steam engine boilers were modelled.

The pressures carried within the boilers of steam engines in the early years were so low that it was deemed unnecessary to 'stay' the flat surfaces, which appeared in the approximately rectangular flat fireboxes. Stays are screwed rods or tubes, provided to support the flat surfaces of a boiler against the bursting effect of internal pressure. But when pressure was increased and it became impossible to leave these flat surfaces unsupported, a method of preventing the collapse of the firebox, together with the bulging and ultimate explosion of the outer shell, was devised. The metal walls of the firebox, normally called sheets, were spaced by stays.

The original engine named *Rocket* was built in 1829 at the Forth Street Works of Robert Stephenson & Co, in Newcastle-upon-Tyne. *Rocket* was an early 0-2-2 steam engine, which was built for (and won) the Rainhill Trials, held by the Liverpool & Manchester Railway in 1829, to choose the best design to power the railway. (Author's Collection)

It was found from experience that any corrosion is usually hidden so it became normal practice on some railways for the stays to have longitudinal holes, called 'tell-tales', drilled in them, which would leak before the stays become unsafe. The

Fixed length threaded rods (also known as stays) secure the distance between the firebox and the heat generated in it, with the outer shell of the boiler containing the water that gets heated to steam. Under these conditions the stays will eventually rot or corrode and will need to be renewed. To maintain the rigidity of the boiler and to make it secure against the potential very destructive forces within, there can be literally thousands of these stays fitted. Here is a row of new stays that have just been partly screwed into their newly refurbished holes prior to permanent fitment, looking like soldiers on parade. (Author)

'crown-sheet', which is the name given to the top of the firebox, had crown stays fitted to it. As the forward portion of the boiler through which the tubes ran is cylindrical, it required no stays to be fitted. So generally, the design of the boiler seems to have followed that of George Stephenson's *Rocke*t.

An improvement to the standard boiler design was the Belpaire boiler, invented by Alfred Belpaire of Belgium. This type of boiler was built with a greater surface area at the top of the firebox, which greatly improved the transfer of heat and so increased the production of steam. Its rectangular shape, however, did make attaching the firebox to the boiler a much more difficult process, but this was offset by simpler interior bracing of the firebox.

The PRR, was an American 'Class I' railroad in the United States having annual carrier operating revenues of $250 million or more and was founded in 1846. Commonly referred to as the 'Pennsy', the PRR was headquartered in Philadelphia. It was the largest railroad by traffic and revenue in the US. for the first half of the twentieth century. Over the years, it acquired, merged with or owned part of at least 800 other rail lines and companies. At the end of 1925, it operated 10,515 miles of rail line; in the 1920s, it carried about three times the traffic of other railroads of comparable length, such as Union Pacific Railroad or the ATSF. The only rival was the NYC, which carried around three-quarters of PRR's ton-miles. At one time, the PRR was the largest publicly traded corporation in the world, with a budget larger than that of the US government and a workforce of about 250,000 people.

The Wootten firebox was very wide to allow the combustion of anthracite coal waste, known as Culm, and was provided with two firing openings. Its size necessitated unusual placement of the crew, examples being the Camelback engine. The Wootten firebox made for a free-steaming, powerful engine and the cheap fuel that was burned was almost smokeless. (Author's Collection)

The PPR used Belpaire fireboxes on nearly all of its steam engines. The distinctive square shape practically became a Pennsy trademark, as no other American railroads except the Great Northern Railway (US) used Belpaire fireboxes in significant numbers. Normally, the top of the firebox was semicircular to match the contour of the boiler, however the Belpaire fireboxes had a squarer shape, which was produced as large as possible within the loading gauge, to offer the greatest heating surface where the fire is the hottest. But the two principal disadvantages of this design were the sharp corners formed at the top of the side plates and the ends of the crown plate, along with the point where the upper portions of side sheets run into the curved portion of the boiler shell.

The use of the 'Belpaire boiler', however, decreased in favour of the 'radial-stay' boiler. In this system, the firebox crown and the outer shell above it, are curved and 'through stays' connect them together. The stays, which are threaded at each end, are screwed into both sheets in the same manner as stay bolts and are placed so as to give the greatest bearing surface to the threads in each plate. The sides of the firebox had in all cases been tied to the outer shell by stay bolts, with their size being dependent upon the forces so incurred. With regard to the shape of the firebox, for many years the lowest portion of the sides was 'narrowed in', to permit the firebox section of the boiler to be dropped in between the frames of the engine. The Wootten boiler, however, which was considered at the time to be a very great advance in design, had a widened out firebox section and was placed on the frames hanging out over the wheels.

But more than that the Wootten firebox was built very wide to enable the successful combustion of anthracite coal waste, known as culm. The size of the firebox was so large that it necessitated the very unusual placement of the crew, with the fireman being at the tender as usual but with the driver re-located to a cab placed on the boiler forward of the firebox, as is seen in such examples as 'Camelback' engines. But with all this aside, the Wootten firebox made for a free-steaming, powerful engine and using the cheap fuel, which burned almost smokelessly, the combination made for an excellent passenger engine and so many 'Camelback' engines operated in passenger service in north America.

The methods of staying narrow fireboxes were equally applicable for those of broad fireboxes and it is evident that these designs could not be carried out without the use of an enormous number of stays. Anything up to 1,400 or even 2,000 stays may have been required for supporting the firebox and crown plates. When high pressures were involved in boilers, it then became necessary for stays to be placed 4 inches or less from their centres. However the stays consequently filled up a large area of the 'water-legs', that is the space between the crown plate and the combustion chamber's end plates.

When 'bad water' was used, the stays became encrusted with lime scale, which prevented proper water circulation and caused unequal contraction of the boiler plates and therefore the more the stays would break. As an example of the trouble associated with stays, G.W. West, the Superintendent of Motive Power of the New York, Ontario & Western Railroad at a discussion of the Wootten boiler design, at the New York Railway Club, 20 April 1899, said 'We have sixty-seven engines on the Ontario & Western fitted with the wide fireboxes. About forty of those are 9 years old, having been built in 1890. Those sixty-seven engines broke 2,703 stay-bolts in 1898. The average for all the engines was twenty-three and a half per engine. The highest broken in any one engine was 120. But the total number of the bolts broken was less than 3 per cent of the bolts in the boiler.'

In 1887, the New York, Providence & Boston Railroad added a trailing axle to a 4-4-0 steam engine in order to spread its weight over more axles. That same year Hinkley built an experimental 'center-cab' 4-4-2 and the Atchison, Topeka & Santa Fe Railway (ATSF) bought a similar experimental

This Wootten Camelback, N° 592, was retired from service and was donated to the Baltimore & Ohio Railroad Museum in May 1954, where it is seen on display. (Author)

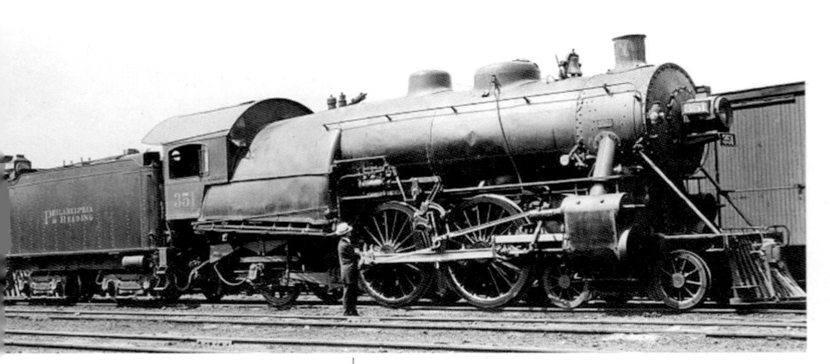

In 1915, the Baldwin Locomotive Works built two 4-4-2 engines with Wootten boilers, for the Philadelphia and Reading Railroad. Here is the first of the two – N° 351. Notice the resemblance to Ivatt's large boiler Atlantics. (Author's Collection)

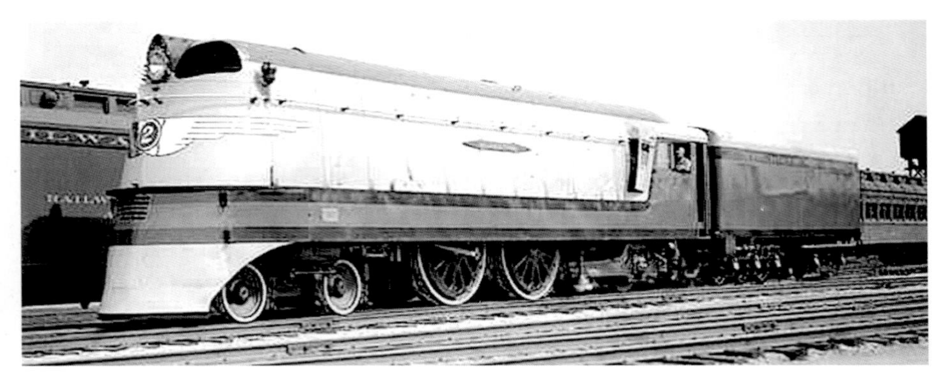

The Milwaukee Road's A class of steam engine comprised four high-speed, streamlined 4-4-2 Atlantic type steam engines, which were built by ALCO between 1935-1937, to haul the Milwaukee Road's *Hiawatha* express passenger trains. They were among the last Atlantic types to be built in the United States and were certainly the largest and most powerful. The class were the first engines in the world to be built for daily operation at over 100mph and were the first class built completely streamlined, bearing their casings for their entire lives. None survive. (Author's Collection)

engine. Around this time, as previously mentioned, the Atlantic Coast Line (ACL) needed an engine with more steaming capacity than their current 4-4-0s, and in 1894 the Baldwin Locomotive Company met the challenge and designed a conventional 4-4-2 engine for the ACL and named the type after them. Other railroads also bought these engines and they too called them Atlantics. However, the Brooks Locomotive Company gave the name 'Chautauqua' to this wheel arrangement and the Chicago, Milwaukee, St Paul & Pacific, The Milwaukee Road, used the name 'Milwaukee' for this wheel arrangement.

At the time with 178 examples, the ATSF owned the most number of engines of this type of wheel arrangement. But the 4-4-2 wheel arrangement was probably made most famous by the Milwaukee Road, when they built four very large streamlined versions of this wheel arrangement to work the Hiawatha service. These engines were the first that were built streamlined and were designed to cruise at 100mph.

The Atlantic City Rail Road (ACRR), 1861-1933, popularly known as the 'Reading Seashore Lines' was a PRR subsidiary, that became the 'Pennsylvania-Reading Seashore Lines' in 1933 and it was the ACRR for which the first engine with the 4-4-2 Atlantic wheel arrangement was built by the Baldwin Locomotive Company in 1895.

The Baldwin Locomotive Company Works

Matthias Baldwin had begun his career as a jewel smith in Philadelphia, but he was more enamoured of technology and in 1828 he built a small engine. It was more than a toy and it powered his first workshop for a period of forty years. This engine now resides in the Smithsonian Institute, Philadelphia, home of the Franklin Institute, which supported new inventions and technology. Philadelphia was also the home of numerous machine shops and soon Baldwin had created a shop where he began to design ever-better steam engines for railroads. He invented the 'leading truck', a pair of wheels that moved along with the engine as the track curved, guiding the engine and so reducing the number of derailments. His idea was later developed and became known as the Jervis truck. His flexible-beam design became very popular because it had more driving wheels and with the weight of the engine over the drivers it could pull longer and heavier freight trains. By 1846, his shop had made forty-two types of engines, allowing him to pay off all his debts and survive the panic of 1837, a severe market reversal. He had also bought out his early partners, making him the sole owner of the company.

Previously in 1831, Baldwin had built, at the request of the Philadelphia Museum, a miniature steam engine for exhibition there. It was such a success that later that year he received an order from a railway company for a full size engine to run on a short line to the suburbs of Philadelphia. The Camden & Amboy Railroad Company and Transportation Company (C&A), the first railroad in New Jersey, had shortly before imported an engine called *John Bull*, from Robert Stephenson & Company, England, which was stored in Bordentown, New Jersey. It had not yet been assembled by Isaac Dripps, under the direction of C&A president Robert L. Stevens, when Baldwin visited the spot. He inspected the detached parts and made notes of the principal dimensions. Aided by these figures, he commenced his task to build his engine to run on a short line to the suburbs of Philadelphia, but more was to follow.

John Bull operated for the first time on 15 September 1831 and it became the oldest operable steam engine in the world, when the Smithsonian Institution operated it in 1981. Built by Robert Stephenson and Co, *John Bull* was initially purchased by and operated for the Camden and Amboy Rail Road, which gave *John Bull* the number 1 and its first name, *Stevens*. The C&A used this engine heavily from 1833, until 1866, when it was removed from active service and placed in storage.

Built by Robert Stephenson and Co, the 0-4-0 engine *John Bull* was initially purchased by and operated for the Camden and Amboy Rail Road and Transportation Company (C&A), the first railroad in New Jersey. Given the number 1 and its first name, *Stevens*, the C&A used the engine heavily from 1833 until 1866. Originally, when the engine had been imported to America in 1831 and while it was still in a kit of parts, Matthias Baldwin inspected the detached components and made copious notes of the principal dimensions. Aided by these figures, Matthias Baldwin commenced his task of building his own design of steam engines, but on a large scale. In some 120 years, Baldwin's built 70,500 of the 175,000 railroad engines that were built in the USA. (Author's Collection)

Matthias William Baldwin, 1795-1866, was an American inventor and machinery manufacturer, who specialised in the production of steam engines. Baldwin's small machine shop, which was established in 1825, grew to become the Baldwin Locomotive Works, one of the largest and most successful engine manufacturing firms in the United States. (Author's Collection)

Baldwin commenced his task of building steam engines on a large scale and in 120 years Baldwin's built 70,500 of the 175,000 locomotives built in the USA. The original Baldwin plant was located on Broad Street in Philadelphia, where the company did business for seventy-one years, until it moved in 1912 to a new plant in Eddystone. Baldwin made its reputation building steam engines for the Pennsylvania Railroad, the Baltimore & Ohio Railroad, the Atchison, Topeka & Santa Fe and many of the other railroads in North America, as well as for overseas railways in England, France, India, Haiti and Egypt. Thereby the Baldwin Locomotive Works of Philadelphia became the most successful builder of steam engines in the world.

In eight weeks, thousands of workers at the Baldwin Works

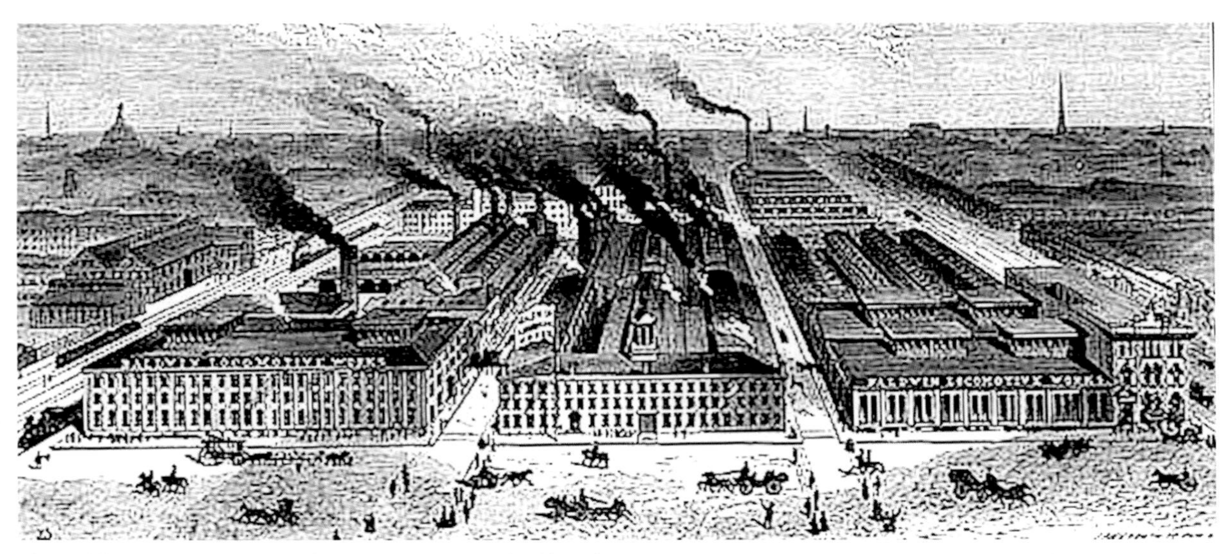

The Baldwin Locomotive Works was an American builder of steam engines that was originally located in Philadelphia, Pennsylvania. The company was a very successful producer of steam engines, but it stopped producing engines in 1956 and went out of business in 1972. (Author's Collection)

would take a custom design from drawings to finished product. By 1905, more than seven engines rolled out of the plant every day. Baldwin was the premier engine manufacturer and ranked at the top of American industry for sixty years. Ironically, the major strength of the Baldwin company, which was the ability to build custom products efficiently to the customer's specific demands, was to prove a major weakness as the economy changed to a more consumer-driven environment where standardised products, hierarchical managerial structures and market control strategies dominated. But by 1906, Baldwin was producing an engine every three hours, twenty-four hours a day. Baldwin's success came because of his reliance on a core of skilled workers rather than on trying to improve profits by manipulating workers and exploiting them. But Baldwin's could not compete with manufacturers who mass-produced standard designs.

By the late 1940s, it was very clear that the steam engines' days were in decline and that each of the big three steam engine builders were far behind EMD, Electro-Motive-Diesel, who were the originators of diesel-electric motive power for the railways. The company was formed in 1922 and supplied the first ever diesel-powered shunters and in 1934, EMD introduced the original lightweight passenger engine for transcontinental use. In 1939, the company pioneered diesel power for mainline freight trains, with many new diesel designs and a growing customer base, it was the way that the world was moving on. Lima Locomotive Works was an American firm that manufactured railroad locomotives from the 1870s through the 1950s, and the company took the most distinctive part of its name from its main shops location in Lima, Ohio. Lima merged in 1947, with General Machinery Corporation of Hamilton, Ohio, to form Lima-Hamilton. Engine builder Hamilton did this in an effort to get a foot hold in the diesel market but made little progress. In desperation Lima-Hamilton merged with Baldwin in 1950 to become the Baldwin-Lima-

In 1905, the Pennsylvanian Railroad company built an engine test plant at Altoona, near 17th Street and required a staff of twenty-six people to operate it. This view inside shows one of the Pennsylvania Rail Road's Atlantic 4-4-2 engines on the rolling road, waiting for tests to commence. (Author's Collection)

Hamilton Corporation (BLH). However, by 1956 BLH ceased production of common carrier size engines.

The Atlantic name became particularly well known when the PRR began using engines of this type on its high speed commuter service between Camden and Atlantic City, when they eventually brought the shortest time over this 55.5 miles rout, down to just fifty minutes start to stop and it became the fastest railway in the world in the first decade of the new century.

The PRR adopted a classification scheme for its engines early on. Initially, each engine type was identified by a letter of the alphabet, starting with Class A and working through the letters. Only standardised types were classified in this way; non-standard, inherited engines were just classed as 'odd' and their wheel arrangement, e.g. 'odd 4-4-0'. The limitations of this scheme became obvious as soon as the railroad passed twenty or so classes of standardised engine. So a new scheme was adopted, which would classify the railroad's steam engines until the end of steam operation in 1957. Each engine's wheel arrangement was assigned a capital letter and then a subsequent number identified the individual type of engine within that overall classification. Subsequent lower-case letters identified variations too small to qualify as a different engine class, generally again counting from 'a' upwards. A lower case 's' immediately following the number stood for a superheated engine. Experimental and acquired engines were normally given higher numbers, in the 1920s and 30s.

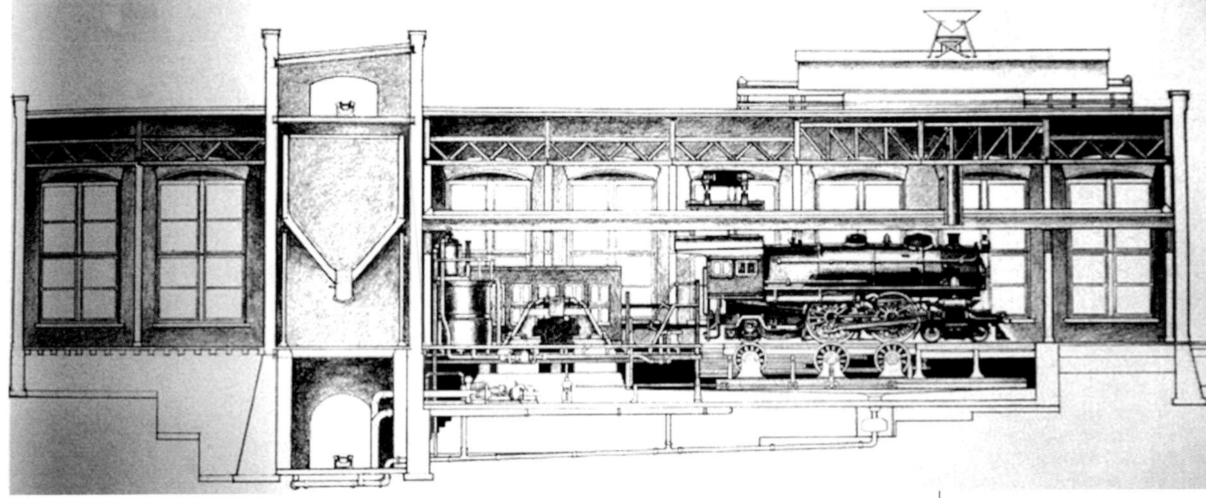

An elevation of the Altoona engine test plant, with an example of the Pennsylvania Rail Road's E6 type Atlantic engine resting on the rolling road. For many years Altoona was the largest railroad shop complex in the world. (Author's Collection)

Electric locomotives were mostly classified under the steam engine classification system and diesel locomotives also had several PRR classifications, but all are better known by the classifications given by their builders.

Class E: 4-4-2 Atlantic

E1: Three of these camelbacks, with Wootten fireboxes, were built in 1899. In addition one E1a was built with a radial-stay firebox.

E2: Eighty-two of these modern (for the time) Atlantics were built in 1901-2. They had radial stay fireboxes, slide valves and Stephenson's valve gear.

E2a: Had a Belpaire firebox; ninety three of these were built between 1902-05.

E2b: Production of this version followed and they had inboard piston valves, with seventy built, both by the PRR themselves and by Alco.

E2c: Covered the classification of Alco-built slide valve equipped engines, of which twenty-two examples were built.

E2d: Thirty-two engines were built by the PRR in Altoona with conventional piston valves and Walscheart's valve gear.

E2s: unsuperheated version of E3.

E3: One hundred and seventy-eight of these Atlantics in various sub-classes were built; many were superheated later on.

E5: Twelve of this class were built and all were later superheated; they were identical to later E3 engines, but for the use of a Kiesel cast steel trailing truck instead of the fabricated truck formerly used.

E6s: Called the 'Apex of the Atlantics', these were some of the last 4-4-2 engines produced in the United States and were the final type of 4-4-2 Atlantic engines built by the railroad. They were Atlantics more powerful than many Pacifics and were excellent 'flatland racers'. Eighty-three were built between 1912-14 and some survived to the end of steam on the PRR in 1957.

E7s: These were rebuilds from class E2 and its subclasses and were given superheating and piston valves. Radial stay E2 engines became E7sa.

E21: Ex-Vandalia class VE1.4 were built by Alco at Schenectady in 1902. They had 20½" x 26" cylinders and 79" driving wheels.

E22: Ex-Vandalia class VE2.5 were built by Alco at Schenectady in 1903. They had 21" x 26" cylinders and 79" driving wheels.

E23: Ex-Vandalia class VE3.10 were built by Alco at Schenectady between 1906-10. They had 21½" x 26" cylinders, 79" driving wheels, a radial stay firebox, inboard piston valves and Stephenson valve gear.

Although Ivatt's small boiler Atlantic N° 990 created a sensation when it first appeared, the impression so created, was mild in comparison with that which greeted his first large boiler Atlantic, in 1902.

N° 251 was an Atlantic engine of American lines in American proportions. It had outside cylinders driving the rear pair of coupled wheels, just like N° 990, but above all with the liberty of development, small diameter trailing wheels were used to carry a Wootten style firebox, spread over the full width of the engine just like its American cousins.

So, using original steam engine concept and design ideas, which had been invented in England, these concepts were developed in North America to become the 4-4-2 Atlantic type engine design. The outcome of this development process arrived back on this side of the 'pond' at the turn of the century and upon its arrival, thoroughly transformed passenger services on the East Coast Main Line, with faster schedules and greater loadings now possible. Interestingly, in 1899 the Midland Railway, the GNR and the GCR all purchased examples of 2-6-0 type steam engines from the Baldwin Locomotive Works in the United States, with the Midland Railway also buying ten similar engines from the Schenectady Locomotive Works at the same time. The GNR's twenty engines were supplied by Burnham, Williams & Co, the proprietors of the Baldwin Locomotive Works and cost £2,418 each, including the tender. The Baldwin 2-6-0 American Moguls were designated as the H1 class on the GNR.

The Works Manager at Doncaster, Douglas Earle Marsh, who was to later become the Locomotive Superintendent of the LBSCR, was sent to Baldwin's in June 1899 to liaise, in the hope of incorporating modifications on these engines for British conditions. This important expedition may well account for the adoption of the wide firebox on both the later GNR Atlantics, the large boiler tender passenger engines and on the large boiler Atlantics that he introduced on the LBSCR. The H1 engines had bar frames, but were not notably large engines: the grate area was only 17.7sq ft, less than that of the later 0-6-0s which had 19sq ft and the total heating surface was 1,380sq ft. Boiler pressure was 175lb psi, Coale pop safety valves and chime whistles were fitted, although it is suggested that British whistles may have been substituted. The outside cylinders were 18" x 24", but as such they were not a novelty at Doncaster. Fay-Richardson balanced slide

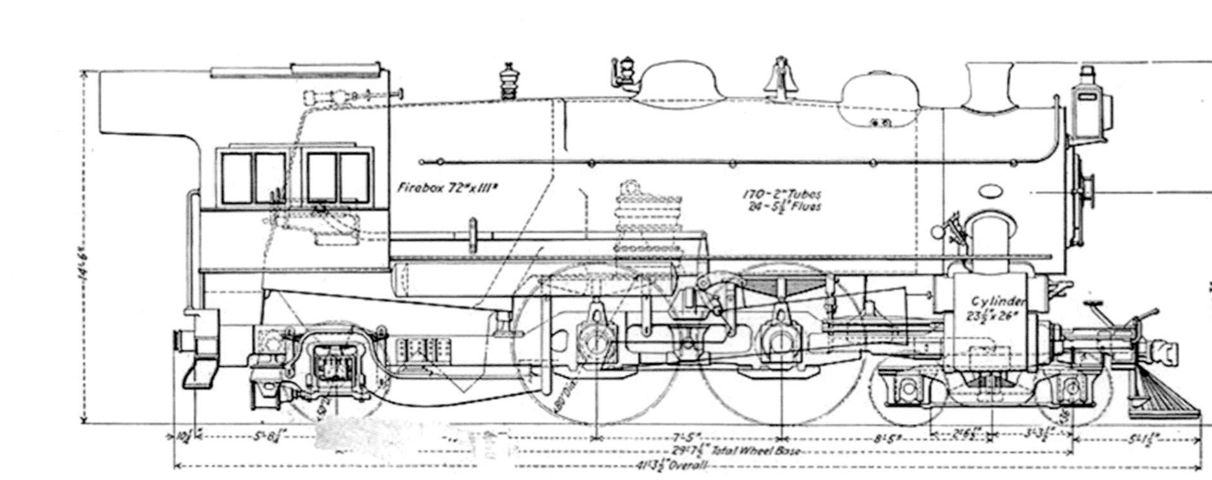

Here is a drawing of the Pennsylvania Rail Road's Atlantic E6 type 4-4-2 engine of 1922. They were considered to be the 'Apex of the Atlantics' and were some of the last 4-4-2 engines produced in the United States. Notice the similarity with Ivatt's large boiler Atlantics. (Author's Collection)

valves were fitted. Their oil consumption was high, but this was not considered to be a problem in the USA, where oil as ever was cheap. Coal consumption at 75lb per mile was also high but this was reduced to 55lb by fitting extended smokeboxes. The American engines were assembled at Ardsley and many worked from there. N° 1200, was exhibited in Vincennes Park at the Paris Expo in 1900.

In 1899, the Great Northern Railway bought twenty steam engines, including the tenders, which were supplied by Burnham, Williams & Co, who were the proprietors of the Baldwin Locomotive Works in Philadelphia, USA, for £2,418 each. The Baldwin 2-6-0 American Moguls were given H1 class designation by the GNR. Here is N° 1182, which was erected from a kit of parts at Ardsley.
(Author's Collection)

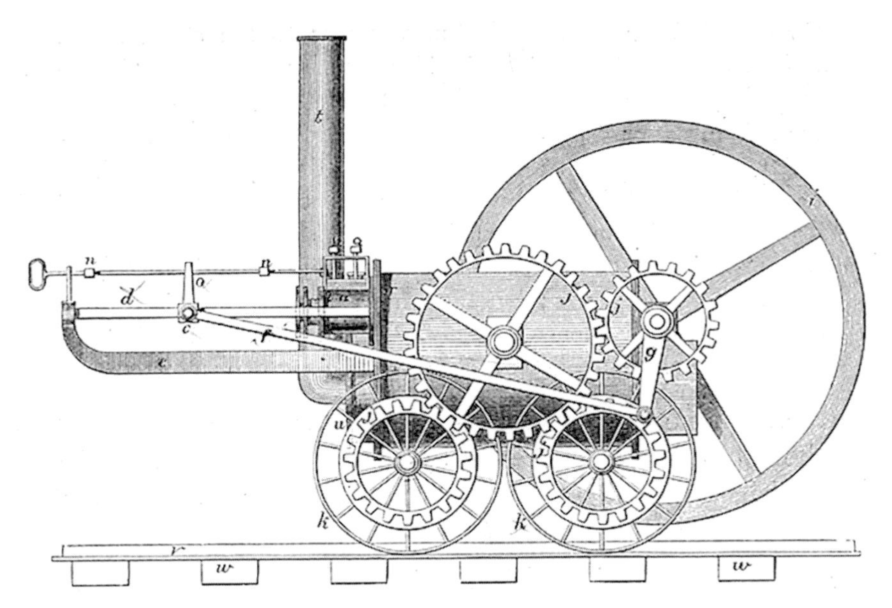

Richard Trevithick (13 April 1771-22 April 1833) was a British inventor, who was born into the mining heartland of Cornwall. Trevithick, who was immersed in mining and engineering from a young age, went on to be an early pioneer of steam-powered road and rail transport. His most significant contribution was the development of the first high-pressure steam engine and the first full-scale working railway steam engine. On 21 February 1804, the world's first engine-worked railway journey took place as Trevithick's unnamed steam engine hauled a train along the tramway of the Penydarren Ironworks in Merthyr Tydfil, Wales. (Author's Collection)

As part of the evolution of steam engine development in this country these 'funny' little American 2-6-0 engines may not have been something special to the railways of Britain in particular; but they may be considered to have been very instrumental in helping D.E. Marsh bringing back with him the design that was important in making the large boiler Atlantic such a revolutionary idea on this country's railways and as such transformed the quality of service for the passenger for the better.

In early 1903, the American 2-6-0s were used experimentally on King's Cross suburban services where they greatly out-performed by 50 per cent the Atlantic 4-4-2T's engines. But they had to be turned and were 'rough riders' and were all

Large boiler Atlantic N° 3297, of the GNR, was completed in April 1905 and lasted until 5 February 1947. It had a superheater fitted during September 1923 and was fitted with piston valves in August 1932. It was built as a direct result of the American Atlantic type engines, which had wide Wootten style boilers fitted and which were renowned for their high speed passenger train working. (Author's Collection)

withdrawn from service between 1909 and 1915.

The 'American Connection' was that 'There is no such thing as an original idea.' Well maybe there is in that the very first steam engine proper, which came into use was Richard Trevithick's engine of 1804. However, human nature being what it is, took these ideas and modified them, they then went to America, were modified again and returned here where they were changed and improved with new ideas. So the 'American Connection' was actually part of the evolution process of the steam locomotive.

Now thats what I call a firebox! The only articulated Camelback type of engine built in North America – indeed there were seven examples built, starting in 1915. These engines, which had Wootten fireboxes, were not suited for high speeds as they had no leading guide wheels. The use of these engines was as very heavy switchers for hump yard work, transfer engines for hauling rakes of wagons between rail yards and as bankers for assisting trains on steep grades. (Author's Collection)

Chapter 5

DONCASTER WORKS

Doncaster's railway works are located in Doncaster, South Yorkshire, and it seems that they have always been referred to as 'the Plant', although no one really knows why. Established by the GNR in 1853, it replaced the previous works located at Boston and Peterborough and only undertook repairs and maintenance until 1867, when a new workshop was built to increase the production and maintenance of engines and rolling stock. In 1866, Patrick Stirling was appointed as the Locomotive Superintendent of the GNR replacing Archibald Sturrock. He decided that he was going to build his own engines rather than buying in 'ready-made' examples, that were really of indifferent quality and did not really do what he wanted them to. His father, Robert, was an engineer, his brother, James, was also a locomotive engineer, his son, Matthew, was Chief Mechanical Engineer of the Hull & Barnsley Railway (HBR) and another son, Patrick, played for Doncaster Rovers FC and was the Mayor of Doncaster.

The first of the 875 class was built in 1886. At this time, the works also began building new coaches, with,

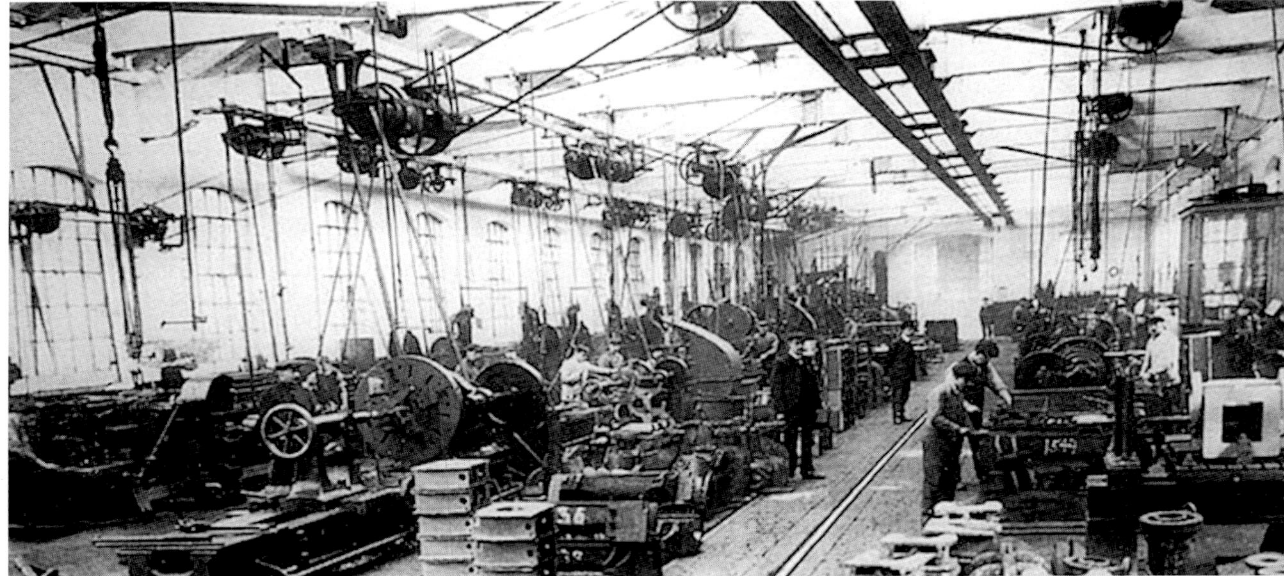

The original railway works located at Doncaster replaced the previous workshops that were located at Boston and Peterborough. Here is an 1896 view of the Doncaster Works machine shop. (Peter N. Townend Collection)

When Patrick Stirling was appointed as the Locomotive Superintendent of the GNR replacing Archibald Sturrock, he wasn't happy buying in 'ready-made' engines and so started building his own. This view shows the boiler shop at Doncaster Works during this period. (Peter N. Townend Collection)

New engines started being built at Doncaster Works in 1866; here is where some of these legendary engines were built. (Peter N. Townend Collection)

in 1873, the first sleeping cars and in 1879 the first dining cars in the UK and in 1882 the first corridor coach. Doncaster Works also constructed various kinds of rolling stock, including wagons and carriages: 99 locomotives, 181 carriages and 1,493 wagons were built in 1891 alone.

Among the engines that the works produced were the GNR's Stirling 4-2-2 engines, otherwise known as the Stirling Singles, the Ivatt 4-4-2s with both small boiler and large boiler versions, the Atlantics.

H.N. Gresley designed his A1 and A3 class 'Pacifics' there, including the world famous *Flying Scotsman* the first steam engine genuinely to achieve 100mph and which also ran from London King's Cross to Edinburgh Waverley non-stop. Another famous class of Gresley's were the A4 class 'Pacifics', including the *Mallard*.

The versatility of the works was such that, like other workshops around the country during the

Between 1905 and 1908, sixty large boiler Atlantic engines were built at Doncaster Works. H.A. Ivatt strongly believed that an engines success depended on its capacity to boil water, so he gave the large boiler Atlantics a large boiler and a large firebox. Although the boiler was large, there wasn't a corresponding increase in cylinder size and the large boiler Atlantics retained the 18.75" x 24" cylinders of the small boiler Atlantic version. (Author's Collection)

Twenty-two examples of Ivatt's small boiler Atlantics were constructed at Doncaster Works. This is N° 983, which was constructed in April 1900 and was given the works number of 875. It was withdrawn from service on 18 April 1936 and was cut up at its birthplace. The small boiler Atlantics were known as Klondikes. (Author's Collection)

Large boiler Atlantic N° 4437 simmers away ready for its next call to duty. Built with slide valves in March 1908, it was withdrawn from active service on 11 March 1944, after which it was put to good use as a stationary boiler at Doncaster Carriage Works, where it was renumbered to become s/b N° 796, from March 1944 to c1965. (Author's Collection)

Whether manufacturing shells or keeping the UK's railways running, the role played by women during the Great War was just as crucial as that of their successors in the Second World War. This view shows woman painting engines, including large boiler Atlantics, in the paint shop at Doncaster Works.
(Peter N. Townend Collection)

A hand-drawn picture shows how a wheel set was changed at Doncaster Works. The engine was moved over the wheel pit, then the lift would descend taking the offending wheel set with it. (Ernie Pay Collection)

Second World War, Doncaster Works joined in the war effort and among other things produced the Airspeed AS.51 'Horsa' aeroplane. This was a troop-carrying glider, which was used for the air assault by British and Allied armed forces during the D-Day airborne assault. It was named after Horsa, the legendary fifth-century conqueror of southern Britain.

In 1948, the LNER was nationalised to form Eastern and North Eastern Regions and part of the Scottish Region of British Railways. The carriage building shop of the works, which had been destroyed by fire in 1940, was rebuilt in 1949, with the production of BR Standard all-steel Mark One coaches in mind. Then, on 16 October 1957, the very last of over 2,000 steam engines that had been built at Doncaster Works, BR Standard 4MT Class 2-6-0 N° 76114,

Here is a general diagram, showing the principal dimensions of the Doncaster designed Pacific engines N° 1470-N° 1480. (Peter N. Townend Collection)

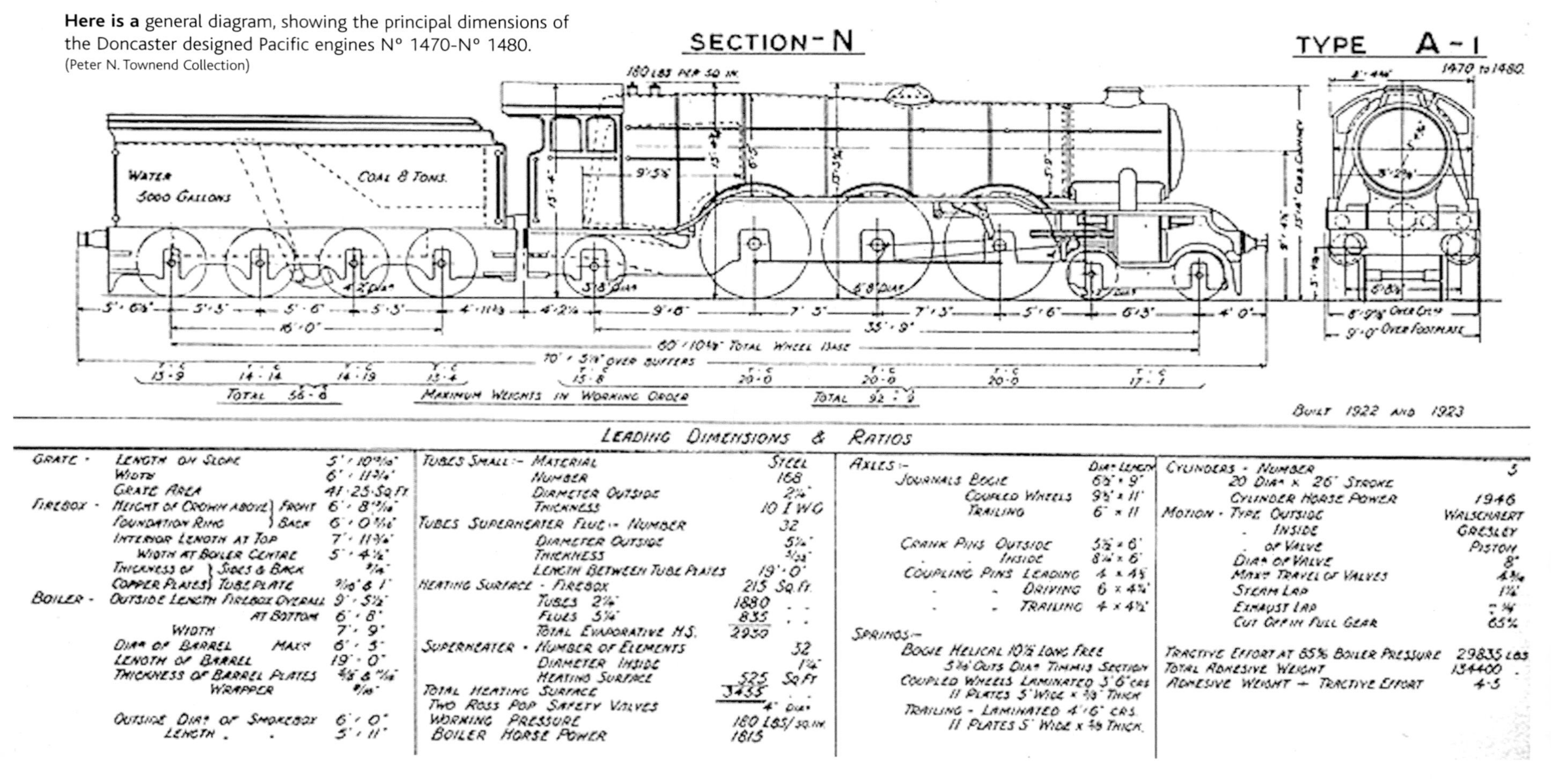

LEADING DIMENSIONS & RATIOS

GRATE -	LENGTH ON SLOPE	5' · 10 9/16"
	WIDTH	6' · 11 3/4"
	GRATE AREA	41 · 25 SQ FT
FIREBOX -	HEIGHT OF CROWN ABOVE FOUNDATION RING } FRONT	6' · 8 13/16"
	} BACK	6' · 0 5/16"
	INTERIOR LENGTH AT TOP	7' · 11 3/4"
	WIDTH AT BOILER CENTRE	5' · 4 1/2"
	THICKNESS OF COPPER PLATES } SIDES & BACK	9/16"
	} TUBE PLATE	9/16" & 1"
BOILER -	OUTSIDE LENGTH FIREBOX OVERALL	9' · 5 1/2"
	AT BOTTOM	6' · 8"
	WIDTH	7' · 9"
	DIAM OF BARREL MAXM	6' · 5"
	LENGTH OF BARREL	19' · 0"
	THICKNESS OF BARREL PLATES	43/64 & 11/16"
	WRAPPER	9/16"
	OUTSIDE DIAM OF SMOKEBOX	6' · 0"
	LENGTH	5' · 11"

TUBES SMALL :-	MATERIAL	STEEL
	NUMBER	168
	DIAMETER OUTSIDE	2 1/4"
	THICKNESS	10 I W G
TUBES SUPERHEATER FLUE :-	NUMBER	32
	DIAMETER OUTSIDE	5 1/4"
	THICKNESS	5/32"
	LENGTH BETWEEN TUBE PLATES	19' · 0"
HEATING SURFACE -	FIREBOX	215 SQ FT
	TUBES 2 1/4"	1880 "
	FLUES 5 1/4"	835 "
	TOTAL EVAPORATIVE H.S.	2930
SUPERHEATER -	NUMBER OF ELEMENTS	32
	DIAMETER INSIDE	1 1/4"
	HEATING SURFACE	525 SQ FT
TOTAL HEATING SURFACE		3455 "
TWO ROSS POP SAFETY VALVES		4" DIAM
WORKING PRESSURE		180 LBS/SQ IN
BOILER HORSE POWER		1815

		DIAM · LENGTH
AXLES :-	JOURNALS BOGIE	6 1/2" × 9"
	COUPLED WHEELS	9 1/2" × 11"
	TRAILING	6" × 11
	CRANK PINS OUTSIDE	5 1/2 × 6"
	" " INSIDE	8 1/4 × 6"
	COUPLING PINS LEADING	4 × 4 5/8
	" " DRIVING	6 × 4 3/4"
	" " TRAILING	4 × 4 1/2"
SPRINGS :-	BOGIE HELICAL 10 1/2 LONG FREE 5 7/8" OUTS DIAM TIMMIS SECTION	
	COUPLED WHEELS LAMINATED 3' 6" CRS 11 PLATES 5" WIDE × 5/8" THICK	
	TRAILING - LAMINATED 4' 6" CRS 11 PLATES 5" WIDE × 5/8 THICK	

CYLINDERS -	NUMBER	3
	20 DIAM × 26" STROKE	
	CYLINDER HORSE POWER	1946
MOTION -	TYPE OUTSIDE	WALSCHAERT
	" INSIDE	GRESLEY
	" OF VALVE	PISTON
	DIAM OF VALVE	8"
	MAXM TRAVEL OF VALVES	4 9/16
	STEAM LAP	1 1/4"
	EXHAUST LAP	-1/4
	CUT OFF IN FULL GEAR	65%
TRACTIVE EFFORT AT 85% BOILER PRESSURE		29835 LBS
TOTAL ADHESIVE WEIGHT		134400 "
ADHESIVE WEIGHT ÷ TRACTIVE EFFORT		4·5

Like other workshops during the Second World War, Doncaster Works joined in the war effort and among other things, it produced Horsa gliders for the D-Day airborne assault. The Airspeed AS.51 Horsa, was a British Second World War troop-carrying glider, built by Airspeed Limited and its subcontractors. (Author's Collection)

left the Plant. In 1958, as a result of part of the British Transport Commission plan of 1955, twenty-four electric locomotives were built for the Kent Coast main line electrification scheme. Numbers originally given were E5000-E5023, but the first locomotive, E5000, was later renumbered to become E5024. In 1962, carriage building finished but the works were modernised with the addition of a diesel locomotive repair shop. Under British Rail Engineering Limited (BREL) new diesel shunters and 25kV electric locomotives were built, plus British Rail Type 56 and Type 58 diesel-electric freight locomotives.

In July 2003, 'The Plant'

The innovative skills and adaptability of the Doncaster Works work force has never been in doubt – moving with the times, as styles and modes of transport changed. For example, the British Rail Class 85 was an electric locomotive built during the early 1960s, as part of BR's policy to develop a standard electric locomotive. Five prototype classes 81-85, were built and evaluated, the result eventually led to the development of the Class 86 locomotive. Forty Class 85 locomotives were built from 1961-64 by BR at Doncaster Works. They were built to work trains on the then newly electrified West Coast Main Line, from Birmingham, to Crewe, Manchester Piccadilly, Liverpool and later Preston. By 1965, the tentacles of electrification had spread south to Euston Station in London. Originally numbered E3056-E3095, they were from 1971 onwards, progressively renumbered into the 85001-85040 series. In this image, class member N° 85031, is seen at the buffer stops at Euston Station. It was withdrawn from service in 1990 and was scrapped by MC Metals, Glasgow. (Author's Collection)

The Class 71 was a type of electric locomotive that was used on the 750 Volts DC third rail system on the Southern Region of British Railways. Twenty-four examples were built at Doncaster Works in 1958, primarily for the Kent Coast main lines. Here, example N° E5015 is seen passing through Hildenborough, working the down 'Golden Arrow' service Pullman train, en route to Dover Marine. (Antony M. Ford Collection)

The Class 56 locomotive is a type of diesel locomotive that was designed for heavy freight work. The fleet was introduced between 1976 and 1984. Here is N° 56034, which was part of the batch that was constructed at Doncaster Works. (Author's Collection)

celebrated its 150th anniversary, with an open weekend, but in early 2008, the main locomotive repair shop, which was built on the Crimpsall, was demolished to make way for housing.

GNR Locomotive Designers at Doncaster Works
1846 - 1848 ~ Benjamin Cubitt
1850 - 1866 ~ Archibald Sturrock
1866 - 1895 ~ Patrick Stirling
1896 - 1911 ~ Henry Alfred Ivatt
1911 - 1922 ~ Herbert Nigel Gresley

The LNER inherited the works at Doncaster, as well as Darlington, Gateshead, Stratford, Gorton, Cowlairs, Inverurie, Dunkinfield, Shildon and York, but continued to use Doncaster as its main workshop for the designing, building and maintenance of steam locomotives.

LNER Locomotive Designers at Doncaster Works
1923 - 1941 ~ Herbert Nigel Gresley, knighted in 1936.
1941 - 1946 ~ Edward Thompson
1947 - 1948 ~ Arthur Henry Peppercorn

This aerial view is of Doncaster Works, known for some time as 'The Plant'. Doncaster Station can be seen on the lower right-hand side of the view. (Author's Collection)

Chapter 6

THE TANK ATLANTICS

Almost every major railroad that operated in North America, in the first half of the nineteenth century, owned and operated engines of the 4-4-0 wheel arrangement. Due to the large numbers produced, the 4-4-0 wheel arrangement became commonly known as the American type. The 4-4-0 engine shown here was constructed in 1862 by Mason and Company of Taunton MA and was given the works number 124. In January 1862, the United States Congress had authorised President Abraham Lincoln to seize control of the telegraph and the railroads for military use during the American Civil War. So, the US Military Railroad – USMRR – was set up by the United States Department of War, as a directive of President Lincoln as a separate agency. Its remit was to operate any rail lines seized by the government and this engine – General Haupt – was duly seized. Eventually it was sold to the Baltimore and Ohio Railroad in 1865. Mason is regarded by some to be the father of the American 4-4-0 type of engine. The 4-4-0 American type tender engines, with the frames extended and with a set of training wheels to carry a bunker for coal – replacing the tender – became the basis of the 4-4-2 Atlantic tank engines. (Author's Collection)

GNR C2 Tanks

The 4-4-2T is the tank equivalent of the 4-4-0 American tender engine, but with the frames extended to allow for a coal bunker behind the cab. This necessitated adding a trailing truck to support the additional weight at the rear end of the locomotive. As such, the tank version of the 4-4-2 wheel arrangement appeared earlier than the tender version.

Shortly after his appointment at Doncaster, H.A. Ivatt, of the GNR, built ten examples of the tank version 4-4-2 engines for the West Riding of Yorkshire. Their success led to the building of fifty more, which also had condensing gear fitted for use on the London suburban services as replacements for the Stirling 0-4-4Ts. So a total of sixty C2 class 4-4-2T engines were built in six batches of ten each, between 1898 and 1907. They were all built at Doncaster and were originally designated by the GNR as C2 class. They were never superheated and had boilers that were interchangeable with J4 class 0-6-0s and there was some variations in the chimney designs between different members of the class.

Although the last fifty engines constructed were originally fitted with condensing apparatus for working through the Metropolitan tunnels in London, the rapid increase in weight of trains on the North London services by 1903 revealed that the C2s were not powerful enough to handle the heaviest turns and so a more powerful engine was designed, the

The Ivatt-designed C12 class 4-4-2Ts had boilers that were interchangeable with his 0-6-0 J4 class, although there were some variations in chimneys between different members of the class. Here is an example of the GNR's 0-6-0 J4 class, N° 1095, which became N° 4095 when it became LNER owned. Built in 1896 by Dubs as their N° 3550, it is seen at Tutbury, near Derby during 1923 and lasted until 1934. (Author's Collection)

0-8-2T, which was given the GNR designation of L1 class. But the 0-8-2Ts didn't prove successful and so Ivatt then produced his N1 class 0-6-2T engine which was found to be much better on the heavier turns than the C2s, which were kept for the lighter ones.

When the C2s were replaced on these duties by the larger Gresley-designed N2 class 0-6-2T engines, the condensing apparatus was removed from the C2s and they were drafted away to work on country routes, with a particulary emphasis on branch line duties.

Ten engines were withdrawn before the Second World War and these engines were not included in the LNER renumbering scheme. Forty-nine engines came into BR stock and they were all withdrawn by 1958.

The tank type version of the 4-4-2 wheel arrangement had 5' 7$^{1}/_{2}$" coupled wheels. The boiler used, which had variations for different members of the class, was Ivatt's standard N° 4 domed boiler, which was used in other classes, namely the D2, E1, J4, J23 and to some extent G1, J9, J16, and J18 engines. Initially a working pressure of 170lb

The first ten engines of the new 4-4-2T tank engines were sent to the Leeds area, with the next twenty examples, 1501-1520, going to London. Here is 1504, of the second batch and which was completed in March 1899, at Enfield c1905, complete with its condensing pipes. (Author's Collection)

N° 4514, originally N° 1514, after it had been transferred away from London, when it had had its condensing apparatus removed and a bigger chimney fitted. The GNR's own C2 class Atlantic tank engines when they became LNER became the C12 class. This engine was withdrawn from Hull Botanic Gardens in April 1955, as British Railways engine N° 67371. (Author's Collection)

psi working pressure was used, but was later upgraded to 175lb psi working pressure before the Grouping in 1923, but the pressure was finally reverted to 170lb psi working pressure, which would have eased the boiler repair bills. The cylinders, which were 17" x 26", together with the boiler pressure of 175lb psi working pressure, gave a tractive effort at 85 per cent of 18,428lb. After building forty engines, Ivatt increased the cylinder diameter to 18", for the final twenty engines with the last one being built in 1907. The first ten engines, 1009, 1010 and 1013-20, were all sent to the Leeds area, with the next twenty, 1501-20, going to London. All but 1501 had the short chimneys and domes fitted for underground working, as well as condensing apparatus. N° 1501 was fitted with a standard chimney and dome and after its condensing gear was removed, was banished to Hatfield.

So in spite of the advent of the more powerful 0-8-2T and 0-6-2T tank engines, the 4-4-2T tank engines remained on the King's Cross services for many years. The LNER reclassified these useful little engines as C12 class and drafted them around the system, where they could be seen in such places as Hatfield, Bradford, Chester and Louth. The last C12 class, formerly C2, Atlantic tank engine was withdrawn in 1958 and none has survived into preservation.

Liveries

When they were originally designated as C2 class engines, they were turned out in a standard two-tone lined green livery with light gold black and brown-shaded characters. Also when newly constructed the C2s, in keeping with other GNR engines, had side chains fitted to the buffer beams and so, to avoid these getting in the way, the numerals were set high and were in small figures. After the chains were removed the characters were centralised, with the engine's number displayed on the front buffer beam in $4^{1}/_{2}$" numerals. Ivatt then increased the character size to 6", but Gresley reverted to the $4^{1}/_{2}$" style, which then became the standard form under his tenure.

The two-tone green livery continued up to the First World War, when austerity measures dictated that plain grey with white lettering shaded with black be substituted instead. In both liveries, the engine's number was painted on the back of the coal bunker plating and was located just above the place where the destination board was fitted. After the war, a speedy reversion to green livery was carried out.

GNR C2 class 4-4-2T Atlantic tank N° 1511 was originally sent to London for use on the Metropolitan District services. After the N2 class 0-6-2Ts were sent to London the C2s had their condensing apparatus removed and they were sent north. It is seen here with this apparatus removed and clearly shows a blanking off panel beneath the chimney at the top part of the smokebox. (Author's Collection)

When the Grouping took place in 1923, a new livery policy for every type of engine was formulated, with the green reserved only for express passenger engines, which meant that the LNER's 4-4-2Ts would lose their bright green livery in favour of black with red lining.

Allocation And Work

The first ten 4-4-2Ts (1009-10 and 1013-20), when completed, were sent to work on their designated duties working local passengers services in the West Riding. The next fifty engines, 1501-50, were sent south for use on the London district for use on inner suburban services.

The inner suburban services operated between the following stations: King's Cross, Moorgate, Alexander Palace, Enfield, Highgate and High Barnet. In addition, the engines also worked coal trains across London via Farringdon and Snow Hill to South London depots such as Hither Green and Norwood. In 1905, the fifty engines then existing, were reallocated to the following depots:

Bradford had numbers 1015-18 and 1020.

Leeds had 1009, 1010, 1013, 1014, 1019 and 1501.

London had 1002-08, 1011-12, 1021-40.

The King's Cross Top Shed Metropolitan Link consisted of thirty-five engines, each of which were double shifted. Hornsey used ten engines in their suburban passenger link: 1028, 1010, 1038, 1021, 1003, 1040, 1015, 1036, 1039 and 1016. The turns changed daily and there was a shed day every fifth day. Empty passenger stock duties and other duties were undertaken by both sheds, as were empty carriage pilot duties at King's Cross.

The ten engines numbered 1541-50, built in 1907, were sent to the London District, where all forty-nine condenser engines stationed in the Metropolitan area were shared between King's Cross and Hornsey depots.

The production of Ivatt's N1 class 0-6-2Ts in 1907 generally didn't affect the C2 class 4-4-2Ts, except for the the larger engines assuming the local heavier duties. So that at King's Cross depot in 1914, the N° 1, Metropolitan Link consisted entirely of N1 class 0-6-2Ts and the N° 2, Metropolitan Link comprised of fifteen 4-4-2Ts working on less arduous duties. At Hornsey depot, the Metropolitan Link comprised five 4-4-2Ts.

In 1920, the fifteen C2 class 4-4-2Ts allocated to the King's Cross N° 2 Metropolitan Link were 1507, 1509, 1512, 1515, 1526-27, 1530, 1533-34, 1536, 1540-41 and 1546-48. All engines were 'twelve day shifted' with three shed days. Their duties comprised local passenger, empty stock workings and south London goods workings. The five Hornsey C2s, 1510, 1516, 1518 and 1537-38, were used on local passenger services and some 'milk churn turns' to and from Finchley.

Not surprisingly, the arrival of Gresley's N2 class 0-6-2T engines ousted the Ivatt-designed 4-4-2Ts on London suburban duties and the allocation of 4-4-2Ts at the end of 1922 was:

Ardsley – Nos 1019, 1528,1539 and 1542.

Five of the GNR's C2 class tank Atlantics were allocated to Hornsey, of which N° 1537 was one. It was used on local passenger and milk churn services to Finchley. Here it is seen fitted with a larger chimney, with its condensing apparatus removed, after it had been transferred away from London to Grantham.
(Author's Collection)

Boston – N°s 1504, 1509-10 and 1518.
Bradford – N°s 1009A, 1013, 1015, 1017-18, 1020, 1536, 1540 and 1543-47,
Colwick – N°s 1511, 1515, 1517, N° 1519-24, 1526 and 1530.
Grantham – N°s 1525, 1527, 1529 and 1533.
Hatfield – N°s 1534, 1537, 1541, 1548 and 1550.
Leeds – N°s 1010, 1014, 1501, 1531-32 and 1535-38.
Lincoln – N°s 1016, 1512 and 1514.
Louth – N°s 1503, 1506 and 1513.
Peterborough – N°s 1502, 1505, 1507-08, 1516, 1521 and 1522.

Originally based in the London area as N° 1517, Ivatt tank Atlantic N° 4517 (after it had been renumbered by the LNER at the Grouping), is seen after it had been allocated to Colwick.
(Author's Collection)

A fine elevation shot of Atlantic tank engine N° 4522, after it had spent some time in London and had been transferred to Peterborough. (Author's Collection)

The withdrawal of the remaining Stirling built 0-4-2 and 0-4-4 tank engines was hastened with the arrival of the displaced Ivatt 4-4-2Ts in the north. Here is N° 4502, on shunting duties; notice its GNR Works plate on the bunker side. (Author's Collection)

This gave a grand total of sixty engines.

The arrival of the displaced 4-4-2Ts enabled the withdrawal of the remaining Stirling-built 0-4-2 and 0-4-4 tank engines to take place.

LNER days

When the C2s became reclassified as C12s at the Grouping, 3000 was added to their running numbers. They continued to work on similar duties until the closure of different branch lines, which caused ten examples, in 1930-2, to be transferred to Hull Botanic Gardens in the North Eastern Area, followed by a further six later. While they were allocated to Hull they were employed as working passenger services, but they were disliked by the former Great Central crews. Under the LNER there was a spate of withdrawals with ten going during 1937-8. During the Second World War, some of the C12s were moved to the Manchester District, some to the former Great Eastern and some went to the former Midland and Great Northern lines. June saw 7396 (its second LNER allocated number and originally 1546) withdrawn, leaving forty-nine examples to enter British railways service.

British Railways

In 1948, withdrawals continued in this order: 7358, 7378, 7370 and 7355, plus three more in 1949: 7377,

The GNR's C2 class engine N° 1529 spent the first part of its life in London, before being transferred to Grantham. Here it is seen with a 'King's +' destination board on its smokebox door. (Author's Collection)

The large boiler Atlantics worked the GN main line for many years and Gresley had not rushed to replace them, using them on heavy Pullman workings into the 1930s. Here is a contemporary view of large boiler Atlantic 4461, working the 'West Ring Pullman'. This was the last large boiler Atlantic to be built, being completed in November 1910 at Doncaster Works and lasted until 31 August 1945, when it was formally withdrawn from service. (Author's Collection)

The London, Tilbury & Southend Railway – LTSR's 4-4-2T Atlantic 80, *Thundersley*, was designed by T. Whitelegg and built by Robert Stephenson in 1909 to haul their heavier commuter passenger trains. In 1911, *Thundersley* was decorated in fine style to celebrate the coronation of King George V, as seen in the image. It was withdrawn from traffic in 1956 and was selected for preservation as part of the National Railway Museum's collection. (Author's Collection)

First built by Baldwin and delivered to the Atlantic Coast Line in 1894, the 4-4-2's or Atlantic tender engines found a home working high-speed passenger services on many railroads in North America. They were built over a period of about twenty-five years to work wooden framed passenger cars. But as steel passenger cars became popular during the 1910s, these 4-4-2s could no longer handle the increasing loads and were replaced by more powerful Pacific types, although a few were still in service until the 1950s. Based on a Pennsylvania Rail Road prototype, this die-cast, O-gauge model steam locomotive and tender are authentically painted and lettered. Puffing smoke, with an illuminated headlight, it has a transformer-activated whistle and bell sound, which all add to the realism. Electrical pickups on the engine and tender help provide smooth running. (Author's Collection)

A few months just before the LYR released their first Atlantic inside-cylinder 4-4-2 engine, the GNR introduced to the United Kingdom an outside-cylinder version of steam engine designed by Ivatt with 990 becoming the first to be completed in May 1898. This image shows a contemporary view of the engine, which was later named *Henry Oakley*. (Author's Collection)

The first member of the new large boiler class of express tender Atlantics was 251, which was completed at Doncaster Works in December 1902. It had similarities to those of the small boiler Atlantic class engines, but there was a slight variation to the main and auxiliary frames at the rear end. The major changes in the boiler and firebox led to dimensions which were exceptional by British standards of the day. The layout of the new boiler gave an increase of 72 per cent in evaporative heating surface. Here is a contemporary view of large boiler Atlantic 288 working hard, pulling its train of GNR rolling stock. (Author's Collection)

The LSWR's 415 class was originally rostered for suburban traffic, but the class was soon displaced to the countryside by Dugald Drummond's M7 class. Most of the class was scrapped around the end of the First World War and further decreases meant that all of them were due to be withdrawn by 1929. But the class was noted for its long service on the Lyme Regis branch line and three members of this long obsolete class were used on this duty until 1962, when suitable replacements became available. One example, 488, has survived and can be found on the Bluebell Railway and was the very engine that the author learned to drive on. The image is of a model commissioned by the author to remind him of the happy memories of this training course. (Author)

The Midland & Great Northern Joint Railway, M&GNR, built three 4-4-2 tank engines at Melton Constable; 41, in 1904, 20 in 1909 and 9 in 1910. They had 160lb boiler pressure and 17¼" x 24" cylinders. 41, as seen here, incorporated a number of parts from some of the well-known M&GNJR Beyer, Peacock 4-4-0 engines. Painted in the attractive M&GNJR ochre livery, the 4-4-2T engines were to be seen running in the Cromer area in the early 1930s. (Author's Collection)

As it says on the picture – 'N° 251 Latest Atlantic Type Express Locomotive', as seen on a contemporary postcard. (Author's Collection)

Platform 9¾ at King's Cross Station is now extremely important to young wizards, as that is where one departs for Hogwarts School via the 'Hogwarts Express'. In response to the worldwide popularity of *Harry Potter*, King's Cross Station has had installed an actual 9¾ sign and a 'half-trolley' lodged into a wall, so that fans can visit the station and take fun pictures. Here is the Hogwarts Railway's logo, as is seen on the sides of a Hogwarts Railway's carriage. (Author)

The large boiler Atlantics were used on prestige and express passenger services up and down the Great Northern main line. This image shows a contemporary image of a large boiler Atlantic working a *Flying Scotsman* service past Hadley Wood. (Author's Collection)

The GSWR was a 5' 3" gauge railway company in Ireland and ran from 1844 until 1924. The GSWR grew by building lines and making a series of takeovers until in the late nineteenth and early twentieth centuries it was the largest of Ireland's 'Big Four' railway networks. At its peak the GSWR consisted of a 1,100mile network, of which 240 miles had double track. The core of the GSWR was the Dublin Kingsbridge-Cork main line; Ireland's 'Premier Line'. The company's headquarters were located at Kingsbridge Station. H.A. Ivatt moved to Ireland and the Great Southern and Western Railway at Inchicore where, in 1882, he was appointed to the post of Locomotive Engineer. In 1895, H.A. Ivatt returned to England and succeeded Patrick Stirling as the Locomotive Superintendent of the Great Northern Railway. (Author's collection)

The GNR's Atlantic express passenger engine design was a very important link in the design and evolution of high-speed rail travel in this country and also the world. Gresley followed on after Ivatt and was inspired by the Wootten firebox design, which was very wide and which made for a free-steaming, powerful engine. This led him to produce his Pacific engine design, which ultimately led on to the world breaking *Mallard*, seen here. (Author's Collection)

The Grand Junction Railway's 2-2-2 N° 49 Columbine was designed by Alexander Allen and was built at Crewe Works in 1845. It was withdrawn from service in 1902 and was set aside for preservation. In 1868, Ivatt, at the age of seventeen, was apprenticed at Crewe Works under John Ramsbottom. He would have learned his skills and trade working on engines such as this. (Author's Collection)

Ivatt's time at the GNR was well publicised when he designed the 4-4-2 Atlantic type engine. His large boiler version proved remarkably adaptable to the demands made of them over their years in service. Here is the group pioneer N° 251, as featured on a contemporary cigarette card of the period. (Author's Collection)

When Ivatt died on 25 October 1923, he left behind a legacy that had transformed the quality of travel for passengers on the GNR. This transformation of engine design inspired other railway companies to do likewise. His Atlantic engines were the forerunner of Gresley's Pacifics, which again revolutionised travel on Britain's railway system. Here we see one of his large boiler Atlantics, waiting at York with a south bound express working. (Author's Collection)

A contemporary image of large boiler Atlantic N° 282 during GNR days. This engine lasted until January 1947. (Author's Collection)

Ivatt's large Atlantic large boiler express passenger engine N° 1436 is seen working a heavy teak bodied train. This engine was one of the few that was cut up at Darlington North Road Works, instead of Doncaster Works, when its time came to go. (Author's Collection)

Another contemporary view of a large boiler Atlantic working a 'Leeds - Bradford Express' with '11-on', but this time during LNER days. (Author's Collection)

The rapid increase in the weight of trains running on the North London services by 1903 revealed that the 4-4-2T C2s were not powerful enough to handle the heaviest turns and so a more powerful engine was designed. But the Ivatt designed 0-8-2Ts, which were given the GNR designation of L1 class, did not prove successful and so Ivatt then produced his N1 class 0-6-2T engine, which was found to be much better on the heavier turns than the C2s, which were then cascaded down for lighter duties. (Author's Collection)

The first N1 class engine, N° 190, appeared in 1907, but proved to be too heavy for use on the Metropolitan Widened Lines and so it was moved to the West Riding. The rest of the class were modified and were built with a slightly longer frame and with the radial wheels set further back. The coal bunker and side tanks were also re-styled, with an overall effect of moving the weight back towards the rear. Although the maximum axle load was still high, the new design changes appear to have been successful and so by 1912, there were fifty-one engines operating in the London area. A final total of fifty-six N1s were completed with condensing gear, except for the four engines which were sent for use in the West Riding. (Author's Collection)

Works plate of 1870, from the fiftieth engine to be constructed at Doncaster Works. It belongs to Patrick Stirling's famous 4-2-2 engine with 8 foot driving wheels, his Stirling Single N° 1. (Author)

Gresley instituted a series of improvements to many of the standard Atlantic design engines. Then, in 1915, he rebuilt 279 as a four-cylinder engine with Walschaerts valve gear and rocking shafts to operate the valves for the inside cylinders, but reverted the engine to a two-cylinder design in 1928. Although given a contemporary appearance with a raised running plate, the cab wasn't fitted with a side window. (Author's Collection)

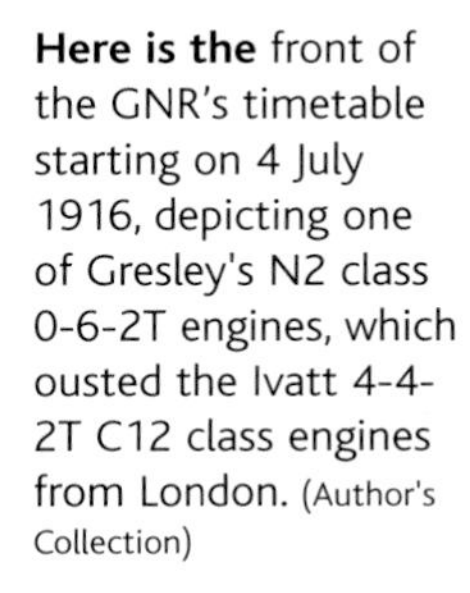

Here is the front of the GNR's timetable starting on 4 July 1916, depicting one of Gresley's N2 class 0-6-2T engines, which ousted the Ivatt 4-4-2T C12 class engines from London. (Author's Collection)

C1 class Atlantic 1421 had brakes that differed from the standard arrangement, by having its 18" diameter brake cylinders for the coupled wheels positioned behind the leading coupled wheels for clearance reasons. 1421, reverted to the normal braking arrangement when it was rebuilt in 1920 as a standard Atlantic. (Author's Collection)

Large boiler Atlantic 284 had a tender, which had three coal rails instead of the usual two. It was common practice to interchange tenders and other engines acquired such tenders after the Grouping. This contemporary image of 284 does however show the tender as having two coal rails. (Author's Collection)

A contemporary view of 1442, when it was specially prepared as the Royal engine. The coat of arms on the rear splashers, brass splasher rims and burnished buffers and wheel rims are clearly seen. (Author's Collection)

Seen passing Brookman's Park in 1937, is C1 class Atlantic 4451, in LNER apple green livery. It was withdrawn from service in May 1948 and was scrapped at Doncaster Works. (K. Leech)

On 11 May 1936, the German built 4-6-4 steam engine 05002 set a new world speed record for steam of 124.25mph. Three members of this class were built in 1935-1937 by Borsig Lokomotiv Werke. 05001 went to the Verkehrsmuseum, Nürnberg, where it can be seen in streamlined condition with its original red livery. The other two engines were scrapped in 1960. (Author)

After *Mallard* had completed her world record-breaking run, she came to rest at Peterborough North. As the engine wasn't able to continue on to London, it was quickly replaced by an Ivatt-designed large boiler Atlantic, 3290, which had originally entered service in June 1904. The ensemble, without *Mallard*, continued onwards to Platform 1, King's Cross and arrived at exactly 18:27, to be met by the press! Here *Mallard* is seen in preservation days with an enthusiast's special working. (Author's Collection)

Seen at Doncaster on 27 September 1953 working the 'Plant Centenarian' are two famous GNR Atlantics, 990 and 251, both looking spick and span in their original colours. They were the centre of an admiring crowd of enthusiasts and were ready to show that even fifty years since being introduced, they were still capable of hauling a train at over 80mph. (Colour Rail Collection)

The GNR's C1 class Atlantic 1442 – the 'Royal' engine – and Stirling's Single N° 1 were both put on display in 1909 at the Imperial International Exhibition at White City, London. Stirling's Single N° 1, went 'Mainline' again 101 years later on 2 June 2010, when it was moved from Southall depot in west London, to Waterloo International via the West London Line, retracing its route and passing the site of the former White City Exhibition en route. N° 1 was being moved to Waterloo to Star in the stage adaptation of E. Nesbitt's The Railway Children at the former International Station's platforms. Here, N° 1, is seen being prepared at the former Southall depot, 81C, for its move to Waterloo. (Author)

On 22 October 2004, Tony Blair – at the time, the Prime Minister and local MP – visited the birth place of railways to officially open a museum called 'Locomotion' – the National Railway Museum's outstation in Shildon, Co Durham. The NRM museum at Shildon is located at the first departure point of the Stockton and Darlington Railway in 1825. The Prime Minister unveiled a plaque with the world famous *Flying Scotsman* beside him. The museum, on the former railway works site where 3,500 people once worked, includes historic buildings as well as a varying collection of some sixty engines. One of those is Sans Pareil which was built by Shildon's most famous son Timothy Hackworth and competed against George Stephenson's Rocket in 1830. Another engine to be seen then was large boiler Atlantic 251, which is seen alongside 4472, *Flying Scotsman*. Both engines contain very little, if anything, of their respective original engines which they are purported to portray! (Stuart T. Matthews)

When the 'Grouping' of railways took place in 1923, the LNER reclassified its C2 class Atlantics as C12 class. Here, N° 7382 (originally N° 1529), transferred away from London to Hull Botanic Gardens depot, is seen with an interesting rake of passenger stock. (Author's Collection)

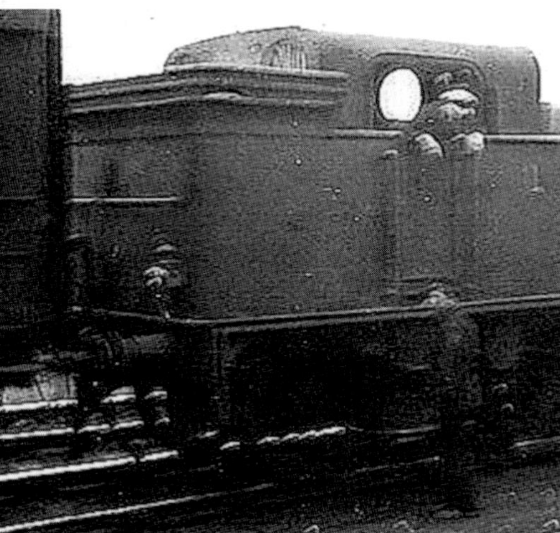

In this view of N° 7382 we see a reverse view of its bunker detail. (Author's Collection)

Seen in the image is Ivatt Atlantic tank N° 4016 (originally N° 1016) of October 1898 and which was part of the first batch of 4-4-2ts constructed, was the fifth from last C12 class 4-4-2T Atlantic tank engine to be withdrawn from service, in March 1948 numbered as N° 7355, never having received its British Railways allocated number of N° 67355. (Author's Collection)

7359 and 7399. From 1953, withdrawals continued, with the last to survive being 67397 (originally the GNR's N° 1547) which, having spent several years of its life at Hull, ended its days at Grantham with it being taken out of service in December 1958 – some fifty-one years of active service. None were saved for preservation.

Seen here is the LNER's C12 class Atlantic tank engine N° 4542 (originally N° 1542) of July 1907, which was numbered as N° 67392 when it was withdrawn from service in October 1956. (Author's Collection)

LNER C12 class N° 4546, seen here, was withdrawn from service in June 1947, never having received its British Railways allocated number of N° 67396. The next C12 class Atlantic tank number in line, N° 67397, was, however, the very last C12 class 4-4-2T Atlantic tank engine to be withdrawn from service, based at Grantham. It was withdrawn in December 1958. (Author's Collection)

Chapter 7

THE SMALL BOILER ATLANTICS – 'KLONDIKES'

The devil is in the detail as they say, so we will now take a closer look at the various parts that made up the the Small Boiler Atlantics, which were affectionately known as 'Klondikes'.

On 22 February 1897, Ivatt submitted a tracing of an experimental express passenger tender engine to the GNR's Locomotive Committee, from whom he received permission to build a working model. Materials were ordered in the following May, even though the drawings were not finalised until December. Twelve months later the first 4-4-2 engine was completed at a total cost of £2,257 11s 6d. The earliest drawing, which was dated February 1897, was for the cylinders. These were 18$^{3}/_{4}$" diameter by 24" stroke and were located outside of the frames and were inclined at '1 in 40.4' from the horizontal, to clear the bogie wheels. The cylinder casting also incorporated the steam chest for the Richardson balanced slide valves, which were arranged vertically between the frames. The steam ports were 1$^{1}/_{4}$", with the exhaust ports at 3$^{1}/_{2}$", by 16". Stephenson link motion actuated the slide valve which had 1$^{1}/_{4}$" lap and a generous maximum travel of 4$^{1}/_{2}$". The engine was fitted with a lever in the cab, to actuate the reverser lever, which was visible between the coupled wheel splashers on the right-hand side. The leading bogie was of the usual standard pattern – no point in reinventing the wheel! It had swing-link suspension, with inside axle boxes. The bottom centre was pivoted 1$^{1}/_{2}$" to the rear of the mid-point of the wheel-base. The four coupled wheels were 6' 7$^{1}/_{2}$" diameter, with the drive from the pistons being transmitted to the rear pair. All of the coupling rods were of I-cross section. An interesting feature was that the connecting and coupling rod pins were turned with the big-end section $^{1}/_{2}$" eccentric to the inner section taking the coupling rod. This differential throw meant that although the piston stroke was 24", the circle traced by the coupling rod travel was only 23". This was done to reduce the stress in these rods at high speed. This practice had been used by William Stroudley some thirty years earlier, when he had modified two engines on the Highland Railway. Eccentric pins

Ivatt's very first 4-4-2 tender Atlantic engine, 990, emerged from Doncaster Works in the summer of 1898 and was one of twenty-two engines built between 1898-1903. It was based on a wheel arrangement which had already established a firm footing in the USA.
(Author's Collection)

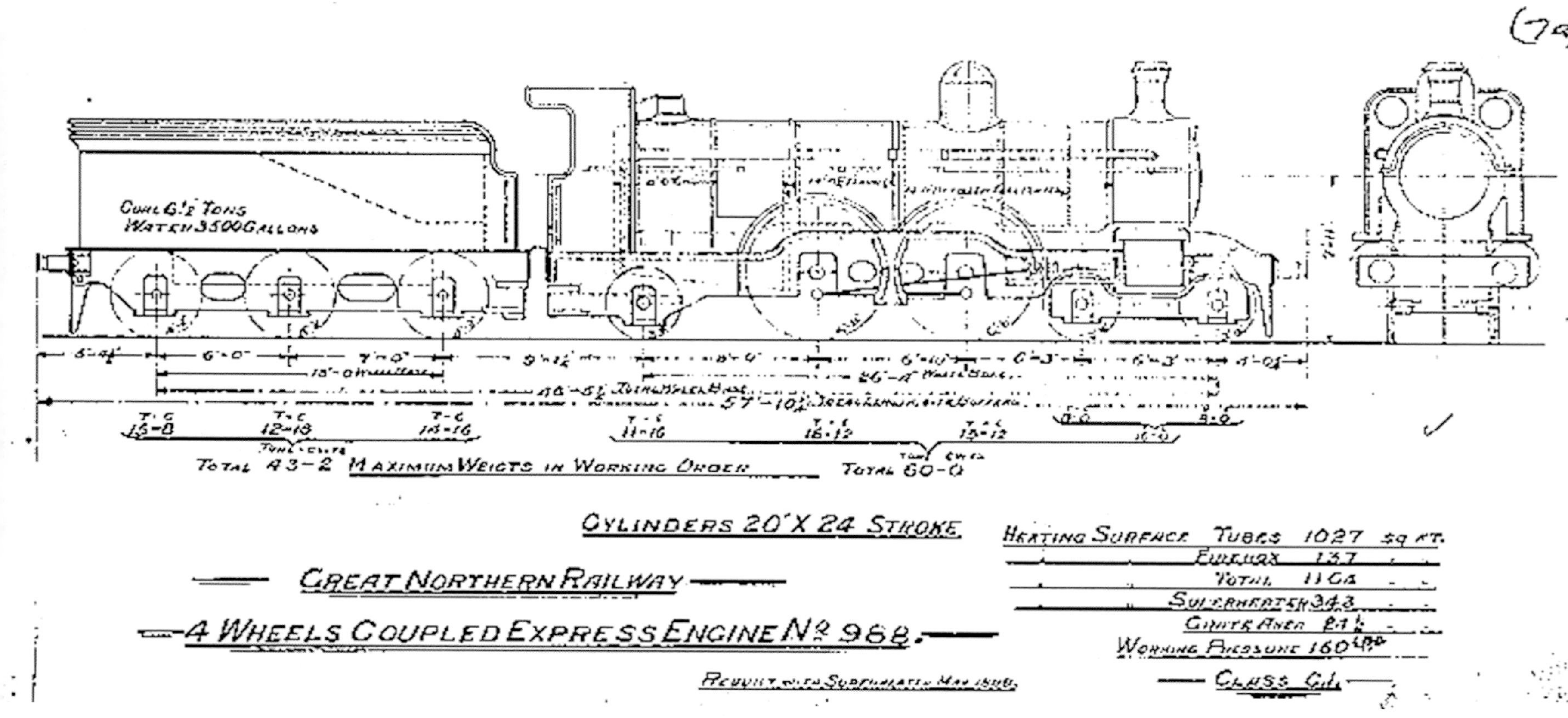

This official drawing shows the general arrangements of Ivatt's small boiler Atlantic tender engines. (Author's Collection)

were also used on LNWR four-cylinder Claughtons and Gresley-designed two-cylinder H2 class 2-6-0 and O1 class 2-8-0 engines.

The frames of N° 990, were 33' 3/4" long overall and were formed from two sections, which were bolted together ahead of the leading coupled wheel horn-guide. The distance between the frames was the usual 4' 1 1/2", but the front-end portion of the frames was 2" closer together to provide extra clearance for the bogie wheels on curves. Separate outside frames 8' 5 3/4" long were provided at the rear end for the trailing carrying wheels with their outside axle-boxes. These frames were 5' 6 1/2" apart with their front edges cut at a slant. The engine was provided with a boiler barrel constructed of three rings, with those at the front and rear overlapping the middle one. The front ring covering the front tube plate had a dummy extension ring riveted to the front of the barrel, which was needed to enable the smokebox to be anchored to the cylinder casting in the usual manner. The front tube-plate was recessed by 1' 11 3/4" and therefore the smokebox outwardly appeared shorter than it actually was.

Two different tube arrangements were originally considered. The first one would have 238 tubes of 1 3/4" diameter, with the distance between the tube-plates 12 feet. In theory, this arrangement would have produced a slightly higher heating surface than the arrangement finally adopted, which had 191 tubes of 13' long and 2" diameter. The 238 tube cross-sectional arrangement was afterwards used in other 4' 8" diameter boiler designs such as D1 and D3 class 4-4-0 engines. The length of the barrel including the extension ring was 14' 8 5/8". 1 3/4" wide water legs above the foundation ring were provided to gain the maximum grate area of 26.79sq ft. However, it was soon revealed in practice that there was a rapid accumulation of sediment, although the detrimental effect was much reduced by the fitting of 3" water legs in the boilers which were produced for subsequent members

In January 1899, after extended trials with the pioneer small boiler Atlantic 990 had finished, twenty further Klondikes were ordered, but shortly afterwards the order was reduced to ten. In this view is 985N, which was one of that batch and which was originally numbered as 985 when it was completed in June 1900. It was renumbered as 985N, from 22 December 1923 and finally became 3985 under the LNER renumbering scheme. It was withdrawn from service in May 1936. (Author's Collection)

of the class. The engine also received a new version boiler during a boiler change between July and November 1903.

It was turned out to traffic in dark shop grey livery on 17 May 1898 and then spent a month on running-in turns from Doncaster. By July it was shedded at Grantham and performed twelve months of regular service before returning to Doncaster on 13 June 1899. Soon afterwards it was fitted with a shelter over the front buffer-beam for taking indicator diagrams and towards the end of the year a speed indicator was fitted to work from the right leading bogie wheel.

In January 1899, after extended trials with N° 990 were completed, twenty further 'Klondikes' were ordered. However, very shortly afterwards, the order was reduced to only ten and so 949-50 and 982-9, were turned out from Doncaster between March and June 1900.

During 1899, even with new additions to the new 'Klondike' class, lengthy preparations took place for building a four-cylinder Atlantic tender engine and, numbered 271, it eventually appeared in July 1902. This was a much more powerful engine in that, although similar in appearance to the previous 'Klondikes', it was provided with four 'high-pressure' cylinders in place of the two carried by the earlier examples. It remained the only engine of its type as, after various modifications to improve it, it was rebuilt and ended up with two inside cylinders only, the form in which it remained until it was scrapped in 1936.

In 1900, when Ivatt had been given the unfamiliar task of naming a Doncaster engine, Webb compounds of his 2-2-2-2 type were among the best known engines in the country and those with double worded names, such as *Queen Empress*, for example, had them spread over two wheel splashers. Ivatt would have been aware of how Webb had done the parallel job of naming a compound engine on the LNWR after the General Manager George Findlay, so what better to do than follow likewise? He certainly used the same split positioning and similar flat brass engraved plates. N° 990 subsequently received the name *Henry Oakley* and was unique in that it was the only GNR engine ever to bear a name until almost the close of that company's independent life, when Gresley's two Pacifics, N° 1470, *Great Northern* and N° 1471, *Sir Frederick Banbury*, which appeared in April 1922, were likewise honoured. The pioneer engine became *Henry Oakley* to honour the Company's retired

This Webb compound engine was sent to the Chicago Exhibition in 1893 – the year that it was built – where it gained the Gold Medal for excellence of workmanship. Subsequently, it ran a train made up of LNWR coaches from Chicago to New York. At the time it was the only British train ever to run in America. It was specially painted white and carried the Royal Arms in honour of Queen Victoria's Diamond Jubilee in 1897. By the end of 1904, this compound passenger engine, *Queen Empress*, had run 473,759 miles. The design and style of nameplates as fitted to *Queen Empress* were copied by Ivatt for the nameplates fitted to *Henry Oakley*. (Author's Collection)

General Manager of 1870-98. His name was divided between the two wheel splashers and only small engraved plates were provided. Their size, style and pairing could well have stemmed from what Ivatt had become accustomed to in his apprenticeship at the LNWR Works at Crewe, under Ramsbottom and Webb. The curved cast brass plates on those two Pacifics set the standard style for LNER locomotive naming until 1935. Throughout its life, *Henry Oakley* remained the only GNR and LNER engine with this distinctive style and pattern of nameplates.

In 1901, the 'Klondike' boiler design was used as the basis for the K1 class 0-8-0 mineral engines and also for the R1 class 0-8-2 tank engines. However, the firebox was 4" shallower at the back so that the ash-pan could be inclined steeply to clear the rear axle. The new boilers, with shallower fireboxes could also be fitted on the 'Klondikes' and the eight-coupled engines.

Around this time it is clear to see that Ivatt had not yet reached his zenith with his Atlantic express locomotive design. His four-cylinder version of the 'Klondikes' – which already had fundamental design and concept differences from the main batch – appeared in January 1902, at exactly the same time as what proved to be the final batch of ten 'Klondikes' was ordered and indeed they were allocated the numbers 251-60. However, by this time Ivatt's ideas had moved on considerably as he had also produced a much larger diameter boiler version which also had a wide firebox. This necessitated that the previous design required an alteration in the main and auxiliary framing at the rear to support it. This all led to a large-boiler Atlantic being ordered in May 1902, whereupon its construction immediately proceeded apace to reach completion during that December. This engine was numbered as N° 251 – which was the next available number – and so it stole the number from the next 'Klondike' that would have been constructed. Instead it became the first engine of a new class and is described in detail in the next chapter.

When N° 251 emerged completed and ready for service, this engine with its much bigger boiler was the largest passenger engine in the country at the time. The essential difference between the new engine and the small boiler Atlantics was the wide firebox, which extended over the whole width of the frames. The large grate was one of the contributory reasons for the success of the design. The wide firebox idea having been adopted from American practice whereby 'wide fireboxes' were seen to be all the rage and were found to be hugely successful. More about the large boiler Atlantics later.

In 1903, N° 251 was followed by the ten previously ordered 'Klondikes' and took the numbers 252-60 and 250. Interestingly, number 254 was the 1,000th locomotive to be built at Doncaster Works. These later engines were constructed with rear end framing similar to that of 251, so that at a later date they could be adapted to take larger boilers if trials with 251 proved successful. In the event however, this was never done, as these engines had inclined screw reversing gear, which tended to complicate and discourage the fitting of a wide firebox. In 1904, 990 had its original reversing lever replaced by screw reversing gear.

The fire-tube superheater was patented by Dr Wilhelm Schmidt in

Specification of GNR 4-4-2 Small 'ATLANTIC' locomotives

	990 class	**990 class**	**990 class**	**Nº 271 as rebuilt**
	Saturated	*Superheated*	*Superheated*	*Superheated*
Motion	Stephenson with slide valves	Stephenson with slide valves	Stephenson with 8" piston valves	Stephenson with 8" piston valves
Cylinders (2)	19" x 24" outside	(a) 20 "x 24" - outside	(a) 20" x 24" - outside	18 ½" x 26" - outside
Boiler:				
Max. dia. outside	4' 8"	4' 8"	4' 8"	4' 8"
Barrel length (b)	14' 8 5/8"	14' 8 5/8"	14' 8 5/8"	16' 0"
Firebox length outside	8' 0"	8' 0"	8' 0"	8' 0"
Pitch	7' 11 ¼"	7' 11 ¼"	7' 11 ¼"	8' 1 ¼"
Grate area (sq. ft.)	24.5 (c)	24.5	24.5	24.5
Tractive effort @ 85%	15,640	17,340	17,340	16,070
Boiler pressure - lb/sq in	175	170 (d)	170 (d)	170
Leading wheels	3' 8" (e)	3' 8"	3' 8"	3' 8"
Couple wheels	6' 8" (e)	6' 8"	6' 8"	6' 8"
Trailing wheels	3' 8" (e)	3' 8"	3' 8"	3' 8"
Length over buffers	57 ' 10 ¼"	57 ' 10 ¼"	57 ' 10 ¼"	57 ' 10 ¼"
Weight - full	58T 0C	60T 0C	60T 0C	58T 13C
Adhesive	31T 0C	32T 4C	32T 4C	33T 8C
Max. axle load	16T 0C	16T 2C	16T 12C	16T 16C
Engine wheelbase	6' 3" + 5' 3" + 6' 10" +8' 0" = 26' 4"	6' 3" + 5' 3" + 6' 10" +8' 0" = 26' 4"	6' 3" + 5' 3 ¼" + 6' 10" + 8' 0" = 26' 4 ¼"	6' 5 ½" + 6' 8" + 6' 10 ½" + 7' 6" = 27' 6"
Water capacity - gallons	3,670	3,500	3,500	3,500
Coal capacity	5T 0C	6T 10C	6T 10C	5T 0C
(a). Originally 18.75 dia.	-	-	-	-
(b). Including extension ring, without it 12 ft 8 5/8 ins.	-	-	-	-
(c). Originally 26.75 sq ft for Nº 990	-	-	-	-
(d). 160 lb until 1918	-	-	-	-
(e). Originally 6ft 7 ins and 3 ft 7 ½ in diameters	-	-	-	-

1900 and was first used on British railways in 1906, with the principle generally adopted by railways world wide. Briefly, the Schmidt superheater comprises a number of steel pipes that pass backwards from the smoke-box through enlarged fire-tubes, returning to the smoke-box in the form of a long 'U' shape. Steam from the boiler passes through these tubes on its way to the cylinders and is heated to such a very high temperature that you get more 'bang for your bucks'! So, late in 1905, Ivatt informed the GNR Board that the American Loco Co had produced a promising engine superheater, which he would like to try in one of his large passenger engines. The cost was £250 each and the expenditure was approved on 3 November. But in the event, this wasn't done. However, in 1906 a German Schmidt-type superheater was obtained, but it was decided to fit it to an 0-8-0 mineral engine, Nº 417. Favourable results resulted in further eight-coupled engines being similarly fitted.

The first passenger engine to be equipped with a superheater on the GNR was the June 1900 built 'Klondike' Nº 988, in May 1909. It was fitted with an eighteen-element Schmidt-type having the flues arrange in three rows and provided with long loop elements. The superheater heating surface was initially shown as 343sq ft, though this figure was later reduced to 254sq ft due to a different method of calculation, which took into account only the area of the elements inside the flue tubes and was also based on the inside

Here is the former Nº 989, seen working hard. It was renumbered by the LNER becoming 3989 at the Grouping. (Author's Collection)

diameter of the elements and not the outside, so you choose your numbers and you take your choice. The boiler was second hand prior to being re-tubed and had previously been fitted to Nº 254 and therefore had a deeper firebox. The total heating surface was put at 1,505.5sq ft, based on the shallow firebox, which was interchangeable between 'Klondike' and the eight-coupled classes.

Owing to earlier lubrication difficulties with slide valves using dry superheated steam, it was imperative to fit piston valves instead, which at that time suffered from frequent steam leakage. It was also common practice at this time to reduce boiler pressure in superheated engines, which of course reduced the calculated tractive effort. Nº 988 had its boiler pressure reduced from 175lb psi to

160lb psi, although it was raised to 170lb psi at a later date. The reduction in pressure was compensated for by fitting 20" diameter cylinders in place of 18³/₄". Apart from Nº 271, superheating wasn't extended to further 'Klondikes' for almost five years.

In October 1913, Gresley compared the advantages of superheated steam engines over saturated steam ones. The former were more powerful, ran more freely and gave greater fuel economy. So he decided to equip all new and existing engines with superheaters to effect further savings. However, there were disadvantages too, as there was a greater tendency to develop hot bearings. This was overcome by the fitting of suitable mechanical lubricators, using balanced slide valves. At the end of 1913, Gresley announced that 103 goods engines and 34 passenger engines of various classes had received superheaters and that the Robinson type was the cheaper and required less maintenance, as compared with the Schmidt type. The only real technical difference between the two types was in the method of attaching the elements to the header unit.

Gresley designed his own twin-tube superheater, which had seventeen elements each passing through two adjacent flue tubes. It was originally proposed to use 102 small tubes of 1³/₄" diameter but at the last minute, ninety-six tubes 1³/₄" diameter were substituted instead. The design was fitted to Nº 950 in January 1916 and then in 1919 to 255, and 989. In the Gresley system there were two headers, one at the top for saturated steam and the other at the bottom of the smokebox, bolted directly to the steam-pipe castings for superheated steam. Instead of two rows of 5¹/₄" diameter flues each containing four small superheater tubes, four rows of 4" diameter flues were employed. Two lengths only of each of the seventeen elements resulted in larger bridge pieces between the tubes without sacrificing heating surface. Although there were only seventeen as against the eighteen elements in the other types, they extended further into the flues, with nearly a 10 per cent increase being achieved by this means. All elements were interchangeable and another advantage was that one could be replaced without disturbing the others. No dampers or draught retarders were used with the twin tube design.

In August 1914, a scheme appeared for a Swindon superheater to be fitted to a 'Klondike' boiler. There would have been twelve flues 5¹/₄" diameter and 151 small tubes x 1³/₄" diameter. No further information is to hand as to whether this system was actually fitted or whether it was for comparison purposes only. Also in August 1914, five sets of twin-tube equipment were ordered. Three boilers with deep fireboxes were drilled to take the equipment before it was decided to fit them to one 'Klondike' (950), three 0-8-0s and one 0-8-2T. As boilers with deep fireboxes couldn't be fitted to eight-coupled engines, it was necessary to adapt another two boilers with shallow fireboxes to complete that particular order. Two of the deep firebox boilers were then put aside and when two further sets of twin-tube equipment were eventually ordered in October 1918, these were used to re-equip them. In the twin-tube type the elements were originally expanded into the headers in a manner of the Robinson type, but in April 1922, the elements were bolted to the header instead on Nº 255. The boiler pressure of Nº 950 was reduced to 160lb psi until December 1918. All remaining 'Klondikes' received Robinson superheaters by 1924 and those with either Schmidt or twin-tube types were altered to the Robinson type after Grouping, between 1927 and 1929.

Difficulties arising with the early superheating applications, including those on Nº 988, led to detail changes. Originally in these sets, a damper box was fitted over the element ends in the smokebox to prevent overheating. The front plate swivelled forward and down through ninety degrees and was connected to a piston of a steam cylinder located on the left-hand side of the smokebox. With the regulator open, the steam circulated through the elements and was also admitted to this cylinder. The front plate was then pulled downwards to allow hot gases to pass through the flue tubes. On closing the regulator, a balance weight inside the smokebox raised the front plate and so sealed off the flue tubes. Alternatively, the mechanism could be opened manually from within the cab by means of rodding which passed through the hollow boiler handrail or alongside the smokebox where there was a hand lever. A drawback in this arrangement was that it allowed gases to condense in the flue tubes and so produce harmful sulphuric acid, which caused excessive pitting of the exterior surfaces of the elements. Also, there were two anti-vacuum valves located out of sight between the frames, which admitted air to the steam chests when the regulator was closed. But as the cold air was drawn in, it cooled the piston-valve liners and so caused faults to develop in them.

Gresley replaced the steam chest anti-vacuum valves by a valve connected to the saturated side of the header and located behind the chimney. As air was drawn into the elements directly, steam was shut off, thereby preventing the elements from overheating. By the time that the air reached the cylinders it was fairly hot and so avoided any undue variation in the steam chest temperature. The damper box was also dispensed with later and Nº 988 was also brought into line in November 1913.

When quantity production of 'Klondikes' commenced in 1900, apart from Nº 271, the engines were built basically alike except for some minor modifications and some experimental fittings.

The remaining engines in the class differed from the prototype Nº 990, in that the main frames were in one piece with the lap joint replaced by a 1" inward set. They were also 4$^{3}/_{8}$" shallower except for the raised portion above the cylinders where the smokebox was attached. Numbers 250 and 252-60 had the outside rear-end supplementary frames arranged for possible reception of wide firebox boilers. These frames were replaced 1" apart and their front edges were vertical.

In the event of a badly cracked frame, this section was replaced, usually on the fracture side only. Such an exchange was a fairly straightforward procedure if it had a separate front section. With a one piece frame, the front portion had to be removed ahead of the set and disposed of, while the rest of the frame had to have the set straightened out. A new front-end was then bolted to the sound rear end in a similar fashion to that of the original frames of Nº 990. From 1918, the depth of the frame above the cylinders where the smokebox was attached was increased by 2$^{1}/_{2}$" whenever the front end was renewed and an additional row of bolts was provided in some cases. As these frames were renewed on one side only, the engines concerned carried a variation in frame depth between the two sides.

On all except the last ten engines constructed, the cab footsteps were attached to a separate support plate. The bottom step was protruded more than the top one giving overall widths of 8' 6" and 8' 3" respectively. Numbers 250 and 252-60, had modified outside frames so the cab footsteps were attached to these instead. The overall width over top and bottom sets was 8', being the same as the running plate. On the latter, a short hand grip was fitted for the assistance of the engine-men, though this was omitted at first on Nº 990.

All 'Klondikes' were built with 6' 7$^{1}/_{2}$" diameter coupled wheels. with a customary 1$^{1}/_{8}$" deep flange. The short-coupled wheel base of 6' 10" allowed an astonishingly small $^{1}/_{4}$" between the rims! Later the thickness of new tyres was increased from 2$^{3}/_{4}$" to 3" and so increased the diameter at the tread to 6' 8". It was therefore necessary to reduce the flange depth to 29/32" and the distance between the adjacent rims became 3/16"! Contrary to the usual practice with four-coupled engines, the leading coupled axle boxes had underslung plate springs, whilst the trailing ones were provided with helical springs. The trailing carrying wheels had outside plate springs above the axle-boxes. Volute springs, originally used for the bogie axle-boxes, were later exchanged for the helical type.

In September 1903, a special bogie was fitted to Nº 988. The frames were outside of the wheels and wagon-type axle-boxes were fitted. Compared with the standard bogie which had axle journal bearing surfaces 9" long and 5$^{3}/_{4}$" diameter, this had 10" and 5" respectively. Two beams 5$^{1}/_{4}$" apart supported the axle-boxes on each side of the bogie, with an inverted plate spring 3' 6" long located in the space between them. The spring was of the same type as that used for the leading coupled wheels. The ends of the spring were supported by hooks, which were themselves suspended from pins between the beams. The special bogie was removed from Nº 988 in April 1909 and the following August it went to Nº 983, where it remained to the end.

The 'Klondikes' were equipped with steam operated sanding gear for forward running only. On Nº 990, as it was originally built, the sand was discharged in front of the leading coupled wheels. The leading

The Klondikes were equipped with steam operated sanding gear for forward running. N° 990, as originally built, discharged sand in front of the leading coupled wheels. As can be seen, the leading sandbox was above the running plate here, but was placed out of sight on the following engines. (Author's Collection)

sandbox was above the running plate on 990, but was out of sight on the following engines. For reverse running gravity fed sanders operated behind the rear coupled wheels. The sandboxes were located above the running plate behind the rear wheel splashers with an extension to meet the front part of the cab. On 990, the height of the sandbox extension was level with the splasher top, but on the remaining engines it was slightly shallower. Later, the sandboxes on all engines were cut down still further and the rearward extension was dispensed with. All engines were latterly fitted with steam-operated sanding apparatus to work in front of both couple-wheels.

All except N° 990, had screw reversing gear, which was inclined at 20 degrees from the horizontal, with the reversing rod itself curving slightly downwards, although 990, did receive screw reversing equipment at an early date to bring it into line with the then current working practice.

The two outside cylinders were 18¾" diameter by 24" stroke and had balanced slide valves. From 1914 onwards, replacement cylinders with slide valves were made 19" diameter for both saturated and superheated engines. Existing cylinders were rebored to not less than this diameter, as the engines passed through the shops. Replacement cylinders with piston valves, which were fitted to certain superheated engines were 20" diameter. The two types of cylinder were distinguishable by eighteen studs for the 19" version and twenty-two studs for the 20" version, which encircled and secured the front cover.

On the cylinder piston of the superheated engines, which had received piston valves and 950, 982 and 986, which received balanced slide valves, tail rods were employed which extended well forward in front of the cylinder covers.

Between February and August 1912, N° 988's firebox was fitted with a 'Hill patent Locomotive Furnace'. This had steam jets which discharged into special conduits below the grate. They acted like injectors, providing an increased supply of pressurised air which was forced through the narrow spaces between the special fire bars. Another interesting feature was that it incorporated a rocking grate.

A total of twenty-one 'Klondike' boilers were constructed between 1898 and 1903. As built they were not interchangeable between the 0-8-0 or 0-8-2T classes as the ash-pan fouled the rear axles, although after the appearance of 250 and 252, N° 253-60, construction of deep fireboxes ceased.

As a pioneer engine N° 990, when new, received a smokebox door with a diameter of 3' 11¾". The later engines were fitted with new standard doors 4' 9" diameter and N° 900 conformed at its first boiler change in November 1903. Its original boiler, complete with smokebox and small door, was afterwards carried by N° 989, from April 1905 to October 1909.

The handrails on the sides of the boiler originally curved round the top of the smokebox front plate. In addition N° 950, at one time had an extra handrail on the smokebox door itself just above the top hinge strap.

The original fireboxes had nineteen rows of vertical roof stays, with three girder bars at the front end. These numbers were altered to twenty and two respectively from 1909 onwards. Located on each side of the firebox were four wash out plugs provided directly opposite each other. Following the 1922 introduction of transverse stays, which were provided additionally between every fourth row of roof stays, the arrangement and pattern of wash out plugs was altered. Only three were then provided on each side of the firebox. They were spaced further apart and were arranged in staggered formation with those on the right hand side further forward than those on the left.

A drawing dated 21 June 1912 shows a 'Notter's Patent' spark arrester as was fitted to N° 252, at King's Cross shed. The blast-pipe and chimney arrangement remained unchanged but a metal plate was placed some 2 feet in front of three quarters of the area of the front tube plate in the barrel extension. The actual method of working is obscure, but in any event it was ineffective and by January 1913 it had been removed and the smokebox had been extended by 1' $2^3/_4$" instead, with the extra volume acting as a form of spark arrester. The remainder of the class, both saturated and superheated, were similarly treated and all was completed by 1922.

On 15 February 1913, N° 256, was fitted with a 'Schleyder's Patent Smoke Consumer'. Here is an extract from the Patent:

> 'Application filed on 28 May 1906. Serial N° 319.093. Be it known that I am Karl Schleyder, a subject of the King of Bohemia and residing at lfilioiiitz, Bohemia, Austrian Empire.
>
> "Smoke Consumer for locomotive and like boilers".
>
> Combustion, which escapes into the smokebox without being thoroughly utilised, is sucked into the fire-box again for re-combustion in such a manner that the development of ash in the fire-box is not only arrested, but, on the contrary, is actively assisted and the generation of steam is accelerated and increased, while coal is used less and smoke avoided. The improvement assists in the products of re-combustion, which are conducted from the smoke-box into the fire-box through the well-known suction-pipe, being mixed with a suitable quantity of air by special means and of the intensity of the suction of these products of combustion and of the air being controlled and regulated corresponding to the chimney.'

The system was not a success.

The 'Klondikes' had Ivatt-type extended cabs with a slightly curved roof top. The cabs length below the usual cut away portion was only 3' $11^1/_2$" inside, but as the foot-plate extended well into the cab the effective length was just under 3 feet. Inside the cab were sight feed lubricators, which supplied oil to the cylinders and valves. Syphon oil-boxes were provided on the splasher tops for the coupled wheel axle-boxes and horn cheeks.

All of the 'Klondikes' were equipped with vacuum brakes. There were two 18" diameter brake cylinders that were located side by side between the frames behind the bogie, actuating the brake blocks at the rear of the coupled wheels. N° 271, differed in having one 21" diameter cylinder between the coupled wheels and actuated the brake blocks unusually at the front of the leading coupled wheels and at the rear of the trailing coupled wheels. All engines had an additional 18" cylinder under the cab for the brakes on the trailing wheels.

The pioneer Atlantic N° 990, was given the first of Ivatt's B class tenders, which were an updated version of the earlier A class variety. They incorporated a 500 gallon capacity well tank between the frames, thus increasing the total water capacity to 3,670 gallons, although water scoops were later provided. The 13' wheelbase was equally spaced and a five ton coal capacity was provided with two coal rails. The weight was 40 tons 18cwt. Other 'Klondikes' (949-50 and 982-9) were provided with similar tenders with front ends modified to match the cab sheets. The brake blocks were applied at the front of the tender wheels instead of the back, as had previously been the case. N° 271 had a smaller version of tender of 3,140 gallons capacity and was fitted with water scoop apparatus. This tender weighed in at 38 tons 10cwt in working order. The brake block, however, acted on the rear of the wheels in the customary manner. Various kinds of second-hand tenders were attached to 250 and

The pioneer small boiler Atlantic, 990, was given the first of Ivatt's B class tenders – which were an updated version of the earlier A class variety. (Author's Collection)

252-60, comprising both Stirling and Ivatt designs. N° 259, was altered on 24 April 1912 to use fuel oil employing Holden's system. An oil tank inside the tender coal space held 600 gallons of the black gold. Reconversion to coal firing came in April 1913.

N° 271

In 1897, serious attention was being given by locomotive engineers to experiments using four-cylinder engines as opposed to two-cylinder ones, in their quest for a better balanced steam engine. This desirable objective caused Ivatt to do likewise and by the end of 1899, plans were underway for such an engine. This was for a four-cylinder version of the 'Klondikes' but with fundamental differences. It was completed at Doncaster in 1902 and was numbered as N° 271. The four-cylinders were 15" diameter by 20" stroke arranged in line and drove on to the leading axle. The connecting and coupling rods were of the fluted type. Stephenson link motion between the frames actuated two $6^{1}/_{2}$" diameter piston valves located above the cylinders. Each valve spindle carried four heads; the inner ones controlled, by inside admission, the port openings for the outside cylinders, whilst the outer ones controlled, by outside admission, the port openings for the inside cylinders. Unlike those on the 'Klondikes', the boiler had two boiler rings instead of three and the thickness of the plate was 5/8" instead of 9/16". The distance between the tube plates was 1' longer at 14' and there were 141 x $2^{1}/_{4}$" diameter tubes instead of 191 x 2" tubes resulting in a reduction of 139.35sq ft in the overall heating surface compared with the standard 'Klondike'. Attached to the front of the boiler was a dummy extension ring 1' $2^{1}/_{8}$" long and the working pressure of the boiler was 175lb psi.

Trial running during 1902, was conducted between Doncaster, Lincoln and Boston. As with other early applications, the piston valves couldn't lift to release trapped water and were also difficult to lubricate properly. Early the following year great thought was given to rebuilding N° 271, with just two outside cylinders, but the idea was dropped in favour of persevering with the original layout. By June 1903, it was decided to replace the piston valves with conventional balanced slide valves, one above each cylinder, at its next major overhaul. Meanwhile, it was also decided to replace the outside Stephenson valve gear with Walschaerts, which became the first on the GNR. During its time in the works between April and September 1904, N° 271, was altered considerably.

In February 1909. N° 271 was required to be re-boilered and was duly done. A new 0-8-0 type boiler was used, which required a longer extension ring to be fitted to the front of the barrel to compensate for the replacement boiler being 1 foot shorter than the original one used on N° 271. But the revised heating surface was the same as that of the standard 'Klondikes', which had shallow fireboxes. N° 271's original boiler had its front ring shortened and it was re-tubed so that its heating surface was standard with other 'Klondike' boilers and it was next carried by N° 254, from March 1910 until November 1914.

After all this modification work, the performance of N° 271 still proved unsatisfactory and it was taken out of traffic between March and July 1911 for rebuilding as a two-cylinder engine at Doncaster Works. The new cylinders were 18¹/₂" diameter by 26" stroke and were located between the frames with 8" diameter piston valves actuated by Stephenson valve gear. The same arrangement of cylinders, motion and bogie was used in the final series of superheated D1 class 4-4-0 engines 51-65, which were at that time being built. As per the D1s, the overall width over the cylinders of N° 271 was such that to accommodate them within the loading gauge slots had to be cut in the frames through which the cylinder walls protruded, with the bulge being protected by cover plates. The standard bogie was replaced with a new one, having a longer wheel base, so that its wheels could clear the bulge in the frames. The distance from the rear bogie wheels to the leading couple wheels was increased by 6¹/₂". The two new cylinders were bored to 19" diameter in 1923, but otherwise lasted for the life of the engine.

The 1909 boiler of N° 271, was changed when it was rebuilt in 1911 and the new one was equipped with an eighteen element Schmidt superheater. A long extension ring of 16' was fitted at the front, primarily because of the lengthening of the wheelbase and the overall length of the barrel. The boiler pressure was initially reduced to 160lb psi but was raised to 170lb psi shortly afterwards.

A close-up view of the Walschaert's valve gear as used on Klondike engine 271, after it had been rebuilt in 1904. It was the first time that Walschaert's valve gear had been used on the GNR. (Author's Collection)

As previously mentioned N° 271's original boiler was afterwards carried by N° 254, until 1914. Then after remaining spare for four years it was put back again in December 1918, after being suitably altered, including the provision of a longer extension ring and Schmidt superheater. The top row of seven small tubes was gradually dispensed with from the superheated boilers from 1921 onwards and N° 271 was so dealt with in March 1923. At first, the anti-vacuum valves, also known as snifting valves, were located low down on the smokebox sides and led directly to the steam chests. Gresley replaced them in November 1915 by a single mushroom-shaped valve sited some distance behind the chimney. When N° 271 was re-boilered in December 1918, the handrails on the sides of the boiler were pitched 3" higher. At the front of the smokebox the handrail curved round horizontally for a short distance and a separate handrail was provided on the smokebox door. This was placed below the top hinge strap, but was later repositioned above it. The boiler handrails were tubular and rodding passed through them to control the blower valve on the right hand side and the superheater element damper on the left-hand side until the latter feature was dispensed with.

The original July 1902 boiler was finally condemned, when the engine, now numbered by the LNER as N° 3271, was again re-boilered in 1925. Its latest replacement boiler had three washout plugs on each side of the firebox instead of four. In 1928,

During a general overhaul of small boiler Atlantic N° 990, which was carried out from 3 April 1900, seperate brass nameplates – each marked with one word only, forming the name, *Henry Oakley* – were fitted to each driving wheel splasher on both sides of the engine. (Roland Kennington Collection)

although the Schmidt superheater was changed for a Robinson type, the heating surface figures remained unchanged. Four washout plugs reappeared on each side of the firebox when the engine was finally re-boilered in 1931.

Unlike the standard 'Klondikes' N° 271, had four steps located behind the rear bogie and in front of the leading coupled wheels instead of half way along the running plate. A slightly longer hand grip was provided just above these footsteps to assist the engine crew. Later the footsteps received an extra backing plate at the top front end, which curved to meet the underside of the running gear. The chimney height was 2' 6" and was of the type fitted to J21 class 0-6-0s and had a height above rail of 13' 4 13/16". The tall dome cover was 13' 5 3/4" above rail level to top.

Forward running steam operated sanding apparatus was provided in front of the leading coupled wheels, whilst gravity sanding was applied to the rear of the trailing coupled wheels for reverse running. Later steam operated sanders were added in front of the trailing coupled wheels. In 1914, mechanical lubricators were fitted on each side of the engine. The left-hand lubricator fed the coupled axle-boxes and the right hand side lubricator fed the cylinders and valves.

Liveries

On completion in the spring of 1898, N° 990 received a coat of light grey livery for its official photograph. The number appeared on the front buffer beam in the higher position, although safety chains had not then been fitted. Lettering and figures were gilt with grey/black right-hand shading. Black bands edged with very thin white lining divided the boiler cladding and ran around the cab profile. Brass beading adorned the splasher rims and oval brass Works plates were positioned ahead of the sandbox on each side of the frames. The tender was finished similarly to the engine. In July, when 990 was reported to be in regular service, it was noted as being painted in dark Works grey livery when shedded at Grantham, being returned to Doncaster the following year. During 1899, 990 was painted standard green with unlined black borders to the cab sides. During a general overhaul from 3 April 1900, separate brass name plates, incised with one word only, were affixed to each driving wheel splasher on both sides of the engine bearing the name *Henry Oakley*. All brass work was polished. Numbers were in light gold block figures shaded red and brown to the righthand side and below, as were the letters GNR on the tender sides. Side and rear sheets of the tender were green, having a black band edged with white forming panels with curved corners. Outside was a surround of holly green. Engine and tender frames, footsteps, motion brackets and valances were brown edged with vermilion and black lines. Buffer beams were vermilion edged with black but this was changed to a black and white border around 1909. At all times smokeboxes and cylinder covers were painted black.

The above standard livery was applied to the class as a whole, but with the absence of brass beading around the wheel splasher rims on numbers 250, 252 and 260-71 and with the temporary exception of engines undergoing works trials. N° 271, built as a four-cylinder engine in July 1902, was noted in September as painted in dark Works grey with gilt block lettering and figures. It seems not to have received the standard green livery until almost a year later. N° 988, the first passenger engine to be superheated in May 1909, was returned to active service painted plain dark Works grey, including the tender and was shedded at Peterborough. N° 988 was observed in April 1911 to be still in dark grey, although it was officially described as black! The engine was in the charge of Driver J. Eagle, who declared that the engine, whilst in the workshop garb, was nicknamed 'darkie'. Not long afterwards its green livery was restored. Under the LNER, the standard GN green was perpetuated, except for the red-brown frames, which became black, edged with red and the holly green on the tender was abolished. From June 1928 however the LNER decided to economise by repainting the 'Klondikes' in black.

When newly sent into traffic, N° 990 worked out from Doncaster shed still in Works grey livery and appeared in London for the first time working a semi fast train from Peterborough. At this time there wasn't a turntable at King's Cross large enough to accommodate the new engine, and so it had to be turned by using the triangle at Dalston, which was reached via Canonbury. The engine's first allocation was to Grantham shed, where it is said that the District Locomotive Superintendent declined to put it into service and kept it in the shed whilst continuing to use the Stirling 8' Singles instead!

When the 'Klondikes' went to work on the heaviest expresses in November 1905, the allocation was:

London: N°s 252-54
Peterborough: N°s 255-6 and 987-8
Grantham: N°s 257-9, 949-50 and 989-90
Doncaster: N°s 250, 260, 271 and 982-6.

In 1909, small Atlantics were regularly used on the 20:15 night sleeping service to Scotland, usually coming from Grantham shed, whilst London engines 252 and 254 were seen on the 17:30 Newcastle Diner and the 19:55 Highland Sleeper respectively. Peterborough's new N° 988, newly superheated and running in shop grey was noted working a long duty, which commenced in the early morning with a train to Doncaster, then through to London with an express from Newcastle due into King's Cross at 13:40 and returning home on the 15:00 Cromer Express. Many of the sharply timed Manchester expresses were also worked to and from London by Grantham 'Klondikes'. Doncaster's N° 984 worked NER stock on an up Scarborough Express c1909.

Comparative trials were held after July 1911, between the only two superheated piston-valve Atlantics to date, namely N° 271, recently rebuilt with two inside cylinders and N° 988. They worked the 13:30 and 17:45 King's Cross-Doncaster services to compare outside and inside cylinder designs. The contest showed no appreciable advantage except that the outside cylinders were more easily accessible for maintenance.

After rebuilding in 1911, N° 988 was regarded by Doncaster shed as one of their best engines. Until August 1914, it was often used 'turn and turn about', with the large Atlantics on through workings to London which had been instituted in 1902. One such duty in August 1913 was the Sunday 16:45 arrival in London and return the next morning with the 11:45 York train. A more usual duty on which N° 271, was seen in May 1914, entailed an arrival at King's Cross on the West Riding express at 13:55 and a return north with the 16:00 and the engine was on the 09:45 Bradford-London service the next day. From 1910, when all of the large Atlantics were in service, they tended to work the best trains, leaving the smaller variety to handle less important duties including a good deal of fast braked goods trains services.

Until 1915, the only change in allocation was that N° 988 moved from Peterborough to Doncaster. Wartime conditions compelled a reduction in train services from 1915, causing loads to rise to 450-500 tons, which the large Atlantics could easily mange on lengthened schedules, although the small Atlantics could not. Therefore, Grantham received additional large Atlantics and transferred their smaller ones to Doncaster and Peterborough, giving Doncaster almost half of the class. From then on the Peterborough engines, while continuing on secondary main line

During December 1922, just before the Grouping and the formation of the LNER, GNR built Klondike 987 was allocated to Peterborough. Seen here renumbered by the LNER as 3987, this engine was withdrawn from service on the 12 August 1936 and was cut up at Doncaster Works. (Author's Collection)

work, also began working over the East Lincolnshire line to Grimsby and the 'loop' to Lincoln where the *Henry Oakley* was a district favourite.

Few records remain of workings during the First World War, but perhaps the most noteable occasion was when N° 982, was put on the up *Flying Scotsman* service at Grantham, loaded to 490 tons. A maximum speed of 77½mph was reached at Essendine and Peterborough 29.1 miles, was covered in thirty-six minutes fifty seconds. An unchecked run on to London ended in a journey time of 135 minutes, some 4 minutes under schedule.

At the end of the War, further changes in allocation took place including four engines going to York in place of 4-4-0s. However, in the final years of the GNR, Ivatt's graceful small boiler Atlantics still put in a lot of good main line work.

In December 1922, the 'Klondikes' were allocated as follows:

London: N°s 252-4, which were normally sub-shedded at Cambridge)
Peterborough: N°s 255-9, 982, 985 and 987-90
York: N°s 950 and 983-4
Doncaster: N°s 250, 260, 271 and 986.

From 1923, under their new owners, the LNER, the small Atlantics continued to work much the same way. The first withdrawals came in 1935, soon after the introduction of the first of Gresley's A4 class 'Pacifics', when N° 3982 of Hitchin and N° 3988 of Peterborough were taken out of service. The last main survivor was an old London favourite N° 3252, which was withdrawn from Retford in July 1945, although N° 3990, *Henry Oakley*, which was withdrawn on 28 October 1937, survives – see separate section.

In December 1922 and just before the Grouping, GNR Klondike 984 was allocated to York depot. It was then fitted with a superheater in May 1924. It is seen here doing the job it was designed to do, working a train of twelve-wheeler vestibule stock. (Author's Collection)

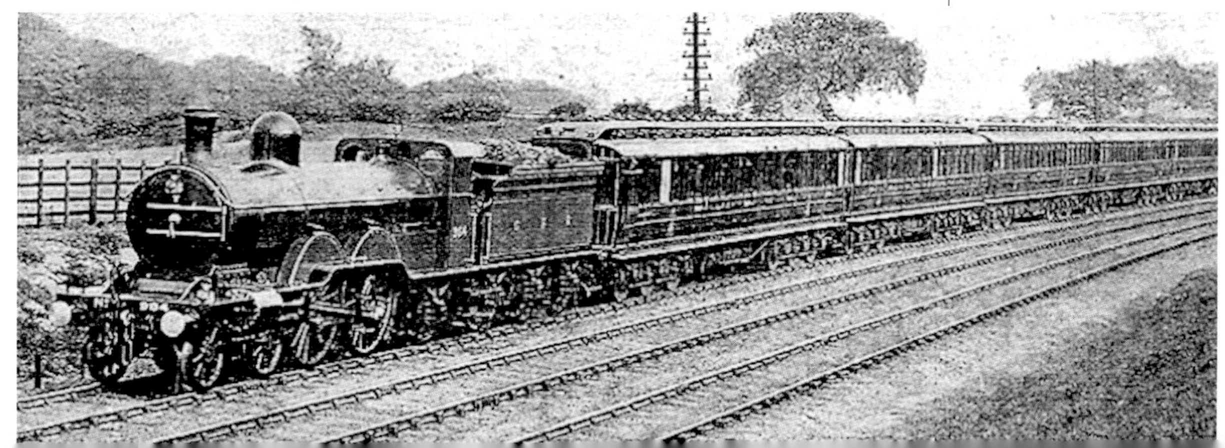

The last Klondike to survive in service was 3252, which was withdrawn from Retford in July 1945. However, 3990, *Henry Oakley* was withdrawn on 28 October 1937 for preservation. It is seen here departing in harness, with a heavy teak bodied train from King's Cross Station in London.
(Author's Collection)

History of GNR 4-4-2 Small 'ATLANTIC' locomotives ~ 1898 - 1947

GNR Number	Date to Traffic	Doncaster Works Number	First LNER Number ~ ( ) = not carried	Second LNER Number ~ ( ) = not carried	Fitted with Superheater	Piston Valves Fitted	Withdrawal Date	Where Cut Up
Nº 250	June 1903	Nº 1003	Nº 3250	(Nº 2894)	August 1918	December 19121	6th June 1944	Doncaster
Nº 252	May 1903	Nº 996	Nº 3252	(Nº 2892)	February 1916	-	14th July 1945	Doncaster
Nº 253	May 1903	Nº 997	Nº 3253	-	March 1916	April 1929	6th May 1937	Doncaster
Nº 254	June 1903	Nº 1000	Nº 3254	-	June 1920	-	2nd April 1943	Doncaster
Nº 255	May 1903	Nº 999	Nº 3255	-	August 1921	August 1924	2nd August 1938	Doncaster
Nº 256	May 1903	Nº 998	Nº 3256	-	October 1918	-	6th March 1943	Doncaster
Nº 257	June 1903	Nº 1001	Nº 3257	-	April 1919	April 1918	26th March 1936	Doncaster
Nº 258	June 1903	Nº 1006	Nº 258N from 16th December 1923, then Nº 3258	-	September 1920	October 1929	21st April 1937	Doncaster
Nº 259	June 1903	Nº 1002	Nº 3259	(Nº 2893)	May 1915	-	2nd October 1943	Doncaster
Nº 260	June 1903	Nº 1005	Nº 3260	-	July 1922	-	28th September 1936	Doncaster
Nº 271	July 1902	Nº 974	Nº 3271	-	July 1920	July 1911	23rd June 1936	Doncaster
Nº 949	March 1900	Nº 872	Nº 3949	-	August 1917	April 1923	19th August 1938	Doncaster
Nº 950	March 1900	Nº 873	Nº 3950	-	August 1923	-	8th December 1937	Doncaster
Nº 982	March 1900	Nº 874	Nº 3982	-	September 1917	-	11th November 1935	Doncaster
Nº 983	April 1900	Nº 875	Nº 983N from 1st September 1923, then Nº 3983	-	March 1918	-	18th April 1936	Doncaster
Nº 984	June 1900	Nº 876	Nº 3984	-	May 1924		29th October 1937	Doncaster
Nº 985	June 1900	Nº 877	Nº 985N from 22nd December 1923, then Nº 3985	-	June 1922	November 1925	11th May 1936	Doncaster
Nº 986	May 1900	Nº 878	Nº 3986	-	March 1918	November 1918	16th October 1937	Doncaster
Nº 987	June 1900	Nº 879	Nº 3987	-	February 1919	February 1918	12th August 1936	Doncaster
Nº 988	June 1900	Nº 880	Nº 3988	-	April 1920	May 1909	17th December 1935	Doncaster
Nº 989	June 1900	Nº 881	Nº 983N from 11th November 1923, then Nº 3989	-	July 1922	-	13th April 1938	Doncaster
Nº 990	May 1898	Nº 769	Nº 3990	-	December 1922	December 1922	8th November 1947	Preserved at the NRM, York

Chapter 8

THE LARGE BOILER ATLANTICS

H.A. Ivatt had held the assumption that an engine's power was derived from its capacity to boil water and so produce available steam in such quantities so as to be useful for the driver to put to good use. So his 'then current trend' towards using larger boilers with increased heating surfaces was not surprising. In December 1902, the same year that small-boiler Atlantic N° 271 which had been provided with four 'high-pressure' cylinders in place of the two carried by the earlier examples appeared, the pioneer of the larger and better known class of Atlantics N° 251 entered service. Its design was based on the 'Klondikes' from the running plate downwards, but with the substitution of a much larger boiler, 5' 6" as against 4' 8" diameter. The most important innovation was a wide firebox extending over the whole width of the frames, giving a much greater capacity to generate abundant quantities of steam. The large grate was one of the contributory reasons for the success of the design. This engine, with its much bigger boiler, was the largest passenger engine in the country at the time.

The basic Atlantic layout turned out to be a most successful design and many technical experiments were carried out to the class during GNR and LNER days. However, it wasn't the first time that an engine with a wide firebox had run on the GNR's lines. In March 1882, a 4-2-2 engine, *Lovett Eames*, manufactured by Baldwin, was brought to Britain to demonstrate a braking system developed by Frederick W. Eames and worked mainly on the GNR. This engine had a firebox devised by J.E. Wootten, which was wide enough to use slow burning small anthracite which was too fine for normal use. There was no direct connection between this engine and Ivatt's designs, other than the fact that Ivatt's engine was inspired by the Wootten type firebox and that the 4-4-2 wheel arrangement was made famous by the Atlantic City engines also built by Baldwins.

Braking on all wheels of a single carriage at once had been accomplished by 1855, but railway engineers dreamed of a brake that could be set on all carriages of a train at once. This 'continuous brake' became the dream of many inventors as well. Numerous schemes were devised using chains, steam, compressed air, vacuum, electricity and even water-pressure. The 'compressed-air' brake was the eventual winner, with George Westinghouse kicking off the race in 1868, when he applied for his first patent. But Westinghouse's brake was less than perfect in its earliest forms and it was expensive and as he improved it and it got more and more expensive.

But there was a competing form of brake that was simple, efficient, reliable and above all cheaper than the Westinghouse air-pressure brake. It required no pump, could maintain full power even with frequent stops, provided a fast application and a quick release. It was the vacuum brake and was, until 1978, the national standard of brake system in Great Britain.

The vacuum brake worked in a very similar way to the early

List of parts 'Head-On' with N° 251, built in 1902.

Compiled by R.H.N. Hardy who was a former shed master, locomotive engineer and a divisional manager at King's Cross.

1. **Stovepipe chimney**
   This particular chimney was temporary, pending final draughting proportions of the well-known GN Ivatt built-up chimney.
2. **Top lamp iron**
   Subsequently deemed to be unnecessary and removed from all engines so fitted within a few years.
3. **Old pattern style of headlamp**
   Replaced by newer pattern shortly after.
4. **Base of lamp iron fitting**
5. **Smoke box door**
6. **Handrail**
   This pattern around the smokebox was never altered. The handrail was hollow to carry the control rod to operate the blower in the smokebox and was operated from lever N° 19 as shown in the cab drawing.
7. **Smoke box door fastening**
8. **The dart handle**
   The dart, which secures the smokebox door, passes through the door centre, turns and engages in the cross bar. When the handle is positioned down vertically, then the dart is in the correct position.
9. **The dart tightening handle**
   The dart is tightened by the use of this handle.
10. **Vacuum train pipe**
11. **Flexible reinforced rubber vacuum pipe**
    This pipe connects with a similar pipe as required.
12. **Cylinder cover**
13. **Buffers**
    These are non-standard, with a tapered shank. Following engines had parallel case buffers as and when, for example, N° 251 after 1923.
14. **Buffer beam**
15. **Side chains**
    These chains were removed when engines were fitted with three-link screw couplings.
16. **Simple three-link coupling**
    These were removed in due course and were replaced with the standard screw coupling.
17. **Richardson balanced slide valve covers**
18. **Position of a dust shield**
    These were fitted to all new Atlantics and those not already fitted.

'straight air' brake, except that in the latter system, air pressure was used to push against the brake mechanisms fitted on each carriage, whereas with the vacuum brake system, a vacuum was created that allowed atmospheric pressure to do the pushing. Instead of a pump on the engine, the vacuum brake had an ejector that sucked air out of the system, allowing atmospheric pressure to push a piston or causing a diaphragm to collapse, this motion in turn was transmitted to the brake gear and caused the train to stop. Vacuum brakes were patented as early as 1844 in England and sixteen years later in the United States. In 1872, John Y. Smith obtained several patents improving the idea and his brake was soon being used on several eastern railroads in America. It provided sufficient competition that George Westinghouse bought Smith's patents in 1875-6 and so produced vacuum brakes of Smith's design for several years under his own name.

About a year after Smith's brake went on the market, Fred W. Eames received his first patent, N° 153,814, dated 4 August 1874. It appears that the primary difference between Eames' brake system and Smith's was that the latter used a piston, mounted on the carriage, while Eames' system used diaphragms mounted separately on each truck. Eames established the Eames Vacuum Brake Company on 14 February 1876, and began manufacturing the brakes in his father's machine shop on Beebee's Island at Watertown, New York. The company was capitalised at $500,000, but was short of capital from the beginning.

This engine was built by the Baldwin Locomotive Works, Philadelphia, to make high-speed passenger runs between New York and Philadelphia. It was equipped with F.W. Eames vacuum brakes and was named Lovett Eames, after the father of the brake's inventor. (Author's Collection)

Within three years, twenty-nine US railroads were using Eames' brake and by 1881, fifty-seven different railroads in nine countries had adopted it. One of its primary users was the New York Elevated Railroad, with light carriages making frequent stops.

Eames travelled extensively marketing his brake. To promote it in England, he purchased a high-speed engine built in 1880 by Baldwin for the PRR and equipped it with his brake. It was shipped to England and in 1881 appeared on a number of lines. But only two of the English lines made purchases, apparently because 'there was already a British-made vacuum brake available' and it was an 'automatic' brake, that is, if the train parted or the ejector failed, the brakes were applied automatically. Eames returned home determined to create his own automatic brake, but found that a New York company with a $47,000 claim against his company had taken over his factory. He sued, but with a mixed result: he would have to pay that company its $47,000, but was given back his factory. In 1885, the Eames Company developed its own automatic brake, spurred on by several railroads that wanted to block the Westinghouse monopoly. In 1886, trials were run on the Burlington railroad. The Eames vacuum brake came out second only to the Westinghouse pneumatic, but it proved too weak for long trains and one of the major goals of the railroads at that time was to increase train length to more than fifty cars.

In 1899, the Midland Railway purchased 2-6-0 freight engines from Baldwin and the Schenectady Locomotive Works. But Samuel Johnson, who was Chief Mechanical Engineer (CME) of the Midland Railway from 1873 to 1903, was critical of the higher (20-25 per cent more in fact) fuel consumption, oil consumption that was 50 per cent greater and repair costs that were 60 per cent greater. The GNR also ordered these 2-6-0s from Baldwin

This 2-6-0 freight engine, 2516, was built for the Midland Railway in 1899, by Baldwin and the Schenectady Locomotive Works. The weight of the tender and engine, complete with coal and water was 86 tons 15cwt.
(Author's Collection)

and gave them the designation of H1 class engines. Indeed Douglas Earle Marsh, the then Chief Assistant to Harry Ivatt on the Great Northern Railway, later to become the Locomotive, Carriage and Wagon Superintendent with the LBSCR, was sent to Philadelphia to liaise with Baldwin's about the finer points of these engines working in this country.

On 10 June 1903 Ivatt brought the following to the attention of the Locomotive Committee: '... the Board is aware of the arrangements to build a further twenty of the 990 class in 1904 and twenty more in 1905. I am anxious to build these with the larger boiler like N° 251. As this will increase dead-weight the Civil Engineer suggests bringing the matter before you as he has informed me that in some of the old structures the factor of safety is now not up to requirements. I want to know as quickly as possible whether these engines are to be built or not as material for some of them must be ordered at once.'

The Committee assured him that necessary work to be undertaken to permit a maximum axle load of 20 tons was possible but bridge stress would need to be checked over an extensive route mileage if present traffic justified introducing heavier engines in large numbers.

On 25 July, Ivatt replied '... the assisting engine mileage increased from 45,696 to 65,280 miles during the present half-year mainly resulting from requests by the Chief Traffic Manager for better timekeeping. Since 1895, mileage rose by 22 per cent though the power of express engines built since then, based on heating surfaces was 20 per cent more than those replaced. As the average weight of

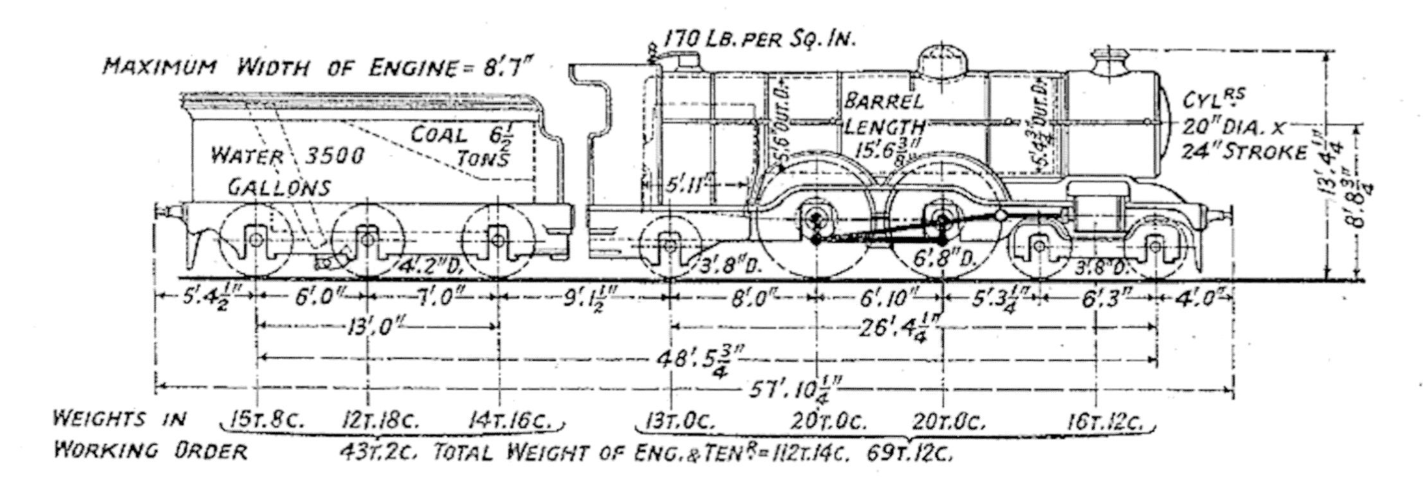

HEATING SURFACE, TUBES—
LARGE AND SMALL ... ... 1,824·0 SQ. FT.
FIREBOX ... ... 141·0 „
TOTAL (EVAPORATIVE) ... ... 1,965·0 „
SUPERHEATER ... ... 568·0 „
COMBINED HEATING SURFACES ... ... 2,533·0 „

SUPERHEATER ELEMENTS ... ... 32–1½ IN. DIA. OUTS.
LARGE TUBES ... ... 32–5¼ IN. DIA. OUTS. } 15 FT. 11¾ IN.
SMALL TUBES ... ... 133–2 IN. DIA. OUTS. } BET. TUBEPLATES
GRATE AREA ... ... 31·0 SQ. FT.
TRACTIVE EFFORT (AT 85 PER CENT. B.P.) ... 17,340·0 LB.

This is an official line drawing of the general arrangements of a Great Northern large boiler Atlantic.
(Peter N. Townend Collection)

these trains had increased by 25 per cent average ton/miles were up by 53 per cent.'

In 1895 several trains were run which couldn't be run by the existing engines, or, as Mr Stirling had expressed it, they were 'fine weather trains'. A remedy sought in the duplication of trains might have been expected, for example in 1896, the 10:00 Leeds and Bradford to London then ran as one 180 ton train from Wakefield; in 1903 it ran as two trains, Leeds at 200 tons and Bradford at 220 tons. In 1896, the 22:30 from King's Cross was one 150 ton train, but then ran as two, the 22:30 at 180 tons and the 22:45 at 200 tons. The increased loads were due to larger rolling stock. Faster schedules were also put into force. Evidently the report was undeniable in its conclusions as it was agreed that more larger engines were required, although by the end of 1903, Ivatt wasn't quite satisfied with the initial performance of the large Atlantic.

N° 251's debut into traffic was marred by a minor design expedient, whereby it had not been possible to provide screw reversing gear to operate at speed and hence this hindered economical working. Valve events were restrictive and there was also a tendency to slip when starting with heavy loads. These disappointing traits in an otherwise excellent engine made Ivatt anxious to test his large Atlantic against contemporary express engine design to effect comparison. The Board shared this feeling and in February 1904 invited tenders from leading manufacturers for the design of such an engine. At the same time, approval was obtained from the Board to build the large boiler versions in considerable numbers.

In his capacity as Ivatt's Works manager and Chief Assistant, Douglas Earle Marsh was closely concerned with the design of the Atlantics and so knew them very well and was aware of their performance etc. After he left the GNR, he then became Locomotive, Carriage and Wagon Superintendent of the LBSCR in January 1905. The following April, he ordered a batch of large boiler tender Atlantic drawings from Doncaster, who were very happy to supply them to Brighton. The principal detail modifications incorporated in the Marsh Atlantics were made in 'red-ink' and were: 26' stroke cylinders; a more spacious cab; a screw reverser assisted by air pressure and a deeper firebox with a 6¹/₄' drop from front to back.

In June 1906, H.A. Ivatt adopted the latter feature for his large boiler Atlantics when he redesigned his wide firebox on similar lines. In this new development the depth at the front end was increased by 3", so that the drop from the front to the back became 6". Firing an Atlantic wasn't so difficult once the fireman learned to ride the swings and jerks at the rear end and this extra slope of the grate tended to make firing easier by shaking the coal forward to the front of the firebox.

Non-articulated compounds rapidly fell out of favour in the early years of the twentieth century in the USA, largely as a result of the practical development of superheating, combined with a move towards faster services, for which many Compounds were ill-suited. Superheating achieved much of the same efficiency as that of compounding, but without complicated equipment. It was noted that when railways failed to keep their more complex Compound types in perfect working order, the engines' overall efficiency sagged significantly. Furthermore, the efficiency gains offered by Compounds in general were offset by disadvantages associated with more complicated operation and more difficult maintenance requirements. An item published in the August 1902 issue of Railway and Locomotive Engineering, USA, discussed the costs of moving freight with Compounds and relayed the opinion of an unnamed railroad president. 'This president, who exercises very great authority and is a close observer of every detail of railroad operating, proceeded to denounce compound locomotives in the most vigorous terms. He did not regard the saving of a few pounds of coal per train mile as being of

Douglas Earle Marsh was born in Norfolk on 4 January 1862 and was educated at Brighton College before going to study at University College, London. He was the Locomotive Superintendent of the London, Brighton and South Coast Railway from 1905 to 1912 and was the man who introduced the Atlantic type of engine to the Brighton Line. (Author's Collection)

List of parts visible in cab of N° 292 built in 1904. Compiled by R.H.N. Hardy who was a former shed master, locomotive engineer and a divisional manager at King's Cross.

1. **Regulator handle**
2. **Duplex vacuum brake gauge**
3. **Steam pressure gauge**
4. **Whistle**
5. **Steam supply to lubricator cock**
6. **Steam supply cock to steam heating**
7. **Sight feed cylinder lubricator**
8. **Steam heating gauge**
9. **LH & RH lifting injectors – Gresham & Craven**
10. **Water regulator**
11. **Warming cock**
12. **Gauge glass protector**
13. **Test cock**
14. **Shut off cock for gauge columns**
15. **Steam supply cock to injectors**
16. **Washout plugs**
17. **Vacuum brake ejector – Gresham & Craven**
18. **Vacuum brake release trigger-chamber side**
19. **Blower cock**
20. **Low pressure cylinders reversing lever**
21. **High pressure cylinders reversing lever**
22. **Operates reducing valve**
    Controlling simple-compound working as do levers 20/21.
23. **Fire-hole door screen**
    Normally open and covers the fire-hole, fire-hole door, trap-door and mouthpiece.
24. **Steam heater safety valve**
25. **Damper control handle**
26. **Compound change-over pressure gauge ie high & low pressures**
27. **Steam sand valve**
28. **Steam manifold controlling 4, 5 & 8**
29. **Drain Pipe from vacuum ejector**
30. **Fireman's seat**
31. **Drag box rubbing plate**
32. **Water supply to LH injector** – the injectors were called lifting injectors, with the steam lifting the water up the pipe.
33. **Water supply to RH injector –** the injectors were called lifting injectors, with the steam lifting the water up the pipe.
34. **Steam heat pipe**
35. **Vacuum brake**
36. **Part of brake mechanism**

any consequence, compared with regularity of service, and he knew that compound locomotives spend much more of their time undergoing repairs than those of simple build. On that account he would have no more compound engines built for his system. When, a compound engine missed a trip undergoing repairs and when it failed on the road, both of which happened frequently, he said, the loss to the company was greater than a gain from coal saving would amount to in a year.'

It appears that the President was not alone in his sentiments. Another issue affecting Compounds involved changes to service requirements as many Compound designs of the era were intended for slow-speed freight services and as railway systems moved on to faster freight services, these slow-speed types became much less desirable.

Ivatt's last development was the fitting of piston valves and Schmidt superheaters to the last ten C1 class Atlantics built in 1910. The addition of a superheater was matched with an increase in cylinder size from $18^{3}/_{4}$' to 20' diameter. At the time, superheating was viewed as a method to decrease the working boiler pressure, which was decreased from 175lb psi to 150lb psi. But due to poor performance, Ivatt decided to build a four-cylinder compound version, N° 292, in March 1905. The outside cylinders drove the trailing coupled axle, whilst the inside cylinders drove the leading coupled axle. Walschaerts valve gear was used on the outside cylinders, but the inside cylinders were fitted with Stephenson valve gear. A valve which was fitted beneath the smokebox switched the engine to operate from simple to compound working. This was controlled from the cab, enabling a driver to switch between simple and compound working. The boiler was built to operate at 225lb psi, but actually operated at 200lb psi. No evidence has been found to suggest that 225lb psi was ever used. Between 1910 and 1912, a standard 175lb psi Atlantic boiler was fitted whilst repairs were being made to the original high pressure boiler.

N° 292, which became the LNER's N° 3292, was never superheated and was withdrawn from service in August 1928. The boiler was equipped with a 24-element Robinson superheater and fitted to the four cylinder simple engine N° 3279.

The GNR directors were not satisfied with the results and so a second four-cylinder compound N° 1300, was built in July 1905. It was built by Vulcan Foundry after the GNR had approached five companies with broad specifications. The layout was similar to N° 292, with the high pressure cylinders on the outside and a divided drive. The boiler had a working pressure of 200lb psi. Due to various troubles in the first few thousand miles, 1300, quickly acquired a bad reputation. Then comparative trials between 1300, 292 and the simple 294, showed that 292, had the best coal consumption, but N° 294 had a lower oil consumption as it only had two cylinders.

In 1914, 1300 was fitted with a Robinson 22-element superheater and the smokebox was lengthened by 9in. The working pressure is believed to have been reduced to 175lb psi, although the engine diagram continued to show a working pressure of 200lb psi. Performance was still unsatisfactory, so when 1300 suffered from a fractured cylinder in 1917, the opportunity was taken to rebuild it as a two-cylinder simple engine. New front-end frames were required, but the non-standard boiler was kept. But 1300, continued to perform poorly and was withdrawn in October 1924, having only run 390,798 miles. This compares to C1 class members 1407-9, which all managed to run over 625,000 miles during the same period.

N° 1421 was the last of Ivatt's C1 class compound engines to be built and it incorporated features from both 292 and 1300. N° 292's layout of cylinders and valve gear was used, but the Stephenson valve gear on the inside was replaced with Walschaerts valve gear. The slide valves were positioned slightly differently, allowing the inside cylinders to be increased in diameter to 18". The boiler pressure was 200lb psi. Although no tests are known about on 1421, it had a reputation of being a better performer than 292, although it also suffered from various problems. Gresley was the first to modify 1421, by adding a 22-element Robinson superheater in 1914. It was rebuilt a second time in 1920, as a standard C1 class Atlantic, with a 24-element Robinson superheater and piston valves.

For comparison, 1300, which was a four-cylinder compound engine that was constructed by the Vulcan

This GNR engine 1421, works number 1166, was built in August 1907 at Doncaster Work. It was withdrawn from service forty years later in August 1947. 1421 was the last of Ivatt's C1 class compound engines and used Walschaerts valve gear. (Author's Collection)

1421 was the last of Ivatt's C1 class compound engines and incorporated features from both 292 and 1300. This is a view of the crank as was fitted in 1907. (Author's Collection)

Foundry, was delivered in July 1905 and differed quite considerably from the standard large Atlantic class in appearance. A further ten standard large Atlantics were built in 1906 and, in 1907, N° 1421 was turned out as Ivatt's last attempt at compounding, incorporating improvements on 292 and 1300.

Construction of standard engines was resumed during 1907-08, when thirty more examples appeared. The fitting of Schmidt superheaters and piston valves, following a similar

Due to poor performance, the GNR directors were not satisfied with Ivatt's four-cylinder compound 292, so a second four-cylinder compound 1300 was built in July 1905. This time it was built by Vulcan Foundry, after the GNR had approached five companies with its broad specifications. The layout was similar to 292, with the high pressure cylinders on the outside and a divided drive. But due to various troubles in the first few thousand miles, 1300 quickly acquired a bad reputation. Comparative trials between 1300, 292 and the simple 296 showed that 292 had the best coal consumption. The image shows 1300, in pristine condition, immediately after it had been built, but before it had had its number painted onto the cab sides. (Author's Collection)

This is an image of the 'Sunny South Express', which was introduced in 1905 and which is formed of LNWR coaching stock being worked by a LBSCR Atlantic tank engine. Part of the service had originated from Liverpool and part from Manchester. The exchange of engines took place between the LNWR and LBSCR at Willesden, where the train was then brought down over the West London Railway to Clapham Junction where it would gain access to the Brighton line. The engine is the famed I3 class Atlantic 23, which featured in the 1909 engine comparison trials on this very service. (Author's Collection)

275, was one of Ivatt's 4-4-2 large boiler Atlantic engines and was constructed in June 1904. It was fitted with slide valves throughout its existence. (Author's Collection)

application to one of the Kondikes in 1909, was the final development under Ivatt on the last ten engines, numbered 1452-61, completed in 1910.

After the Compound interlude, no further technical experiments of note were conducted during 1909. The LNWR were then anxious to test their Precursor class 4-4-0s. One of them, Nº 7, *Titan*, was pitted against one of the LBSCR's 4-4-2T I3 class, with dramatic results in favour of the tank engine. In June of that year Precursors also ran versus GN Atlantics when corresponding trials took place over each company's main line with each type. All engines taking part were saturated, but again the 4-4-0s didn't show up well in terms of fuel economy.

After Gresley succeeded Ivatt in 1911, there were ample main line express engines and so the concentration was directed on goods and mixed traffic types. With construction ceased, the total of Atlantics had reached ninety-four engines and they were numbered 251, 272-301, 1300 and 1400-61. Nº 292, which was built as a four-cylinder compound, was scrapped as such in 1927.

The Pioneer Nº 251 emerged from Doncaster Works in December 1902 having cylinders, motion, wheel diameters and spacings similar to those on the preceding 'Klondikes', but with a variation to the main and auxiliary frames at the rear end. The main change in design was the use of a large boiler fitted with a large firebox, giving dimensions that were at the time quite exceptional by contemporary standards.

The boiler barrel was 15' $6^{3}/_{8}$" long, being formed of two telescopic rings of equal length and of 5' 6" diameter with the dome fitted on the front one. It was pitched at 8' $8^{1}/_{2}$" above rail level.

A very neat looking engine is 1404, following the adage that 'If it looks right, then it probably is'. Having been completed in June 1905, it lasted until 5 August 1947 and was scrapped at Doncaster Works. (Author's Collection)

This image was taken in the 1920s and shows large boiler Atlantic 1440, working a down 'Scotch Express' through Belle Isle in London. The atmosphere of the image is certainly of a different era. Notice the Dewars Whisky sign near the engine's smokebox.
(Author's Collection)

A 1932 view at Barnby Moor as a slow Doncaster-London passenger service is seen being worked by one of Ivatt's large boiler Atlantic engines 4456, which had originally been completed at Doncaster Works in October 1910. 4472 *Flying Scotsman* and its classmates were built to replace this type of engine, due to the demands of increased loadings and speeds. Coincidentally, this picture was taken by Alan Pegler when he was a young schoolboy aged 12 years old and who would ultimately own the world famous *Flying Scotsman.*
(Penny Pegler Collection)

The smokebox length was 5' 9" with an outside diameter of 6 feet. The tube-plate had a $^3/_4$' thickness and was recessed into the front ring, with the distance between the tube-plates being 16 feet. The firebox was 5' 11" long and 6' 9" wide at the foundation ring. The grate was 5' 2" long and 5' 11$^5/_8$" wide with a gradual slope of 3' from the back to the front.

The frames were 33' $^3/_4$" long overall and they were 4' 1$^1/_2$" apart at the rear and 3' 7$^1/_2$" at the front to allow extra clearance for the 3' 7$^1/_2$" diameter bogie wheels on the curves. Separate outside frames 8' 9$^1/_8$" long were provided at the rear end for the trailing wheels and their outside axle-boxes. These frames were 5' 7$^1/_2$" apart.

The foundation ring rested on cast steel brackets, which were bolted to the subsidiary outside frames which were extended and strengthened to accommodate them. To clear the rear coupled wheels, the throat plate and lower section of the firebox tube-plate were set back at an angle.

When first built, the front edges of the firebox cladding on 251 were carried down vertically to the running plate. The tops of the rear driving wheel splashers were continued back on a horizontal plane to meet them. This arrangement did not last long and between 17 January and 3 February 1903, it was modified to the form which became standard. Three washout plugs were provided on the firebox, which together with the boiler barrel and cylinders was lagged with hard wood strips. The spaces between the cylinders and footplate were filled in with insulating material.

As originally built, 251 didn't have a handrail on the smokebox door, but had one fitted below the upper hinge bracket before the engine entered service. The wide firebox prevented the fitting of a screw reverser, so a lever reverser was fitted instead and was located on the righ- hand side of the engine. This entailed enormous levers with eight pins and fulcrum points between the lever in the cab and the suspension links. As with the 'Klondikes', the weights of the expansion links were spring balanced, as were the suspension arms and eccentric rod portions of the valve motion. Experience with lever reverse having notched sectors on the small Atlantic N° 990, showed that a period of service created play on various joints, causing a rattling movement of the lever and strain on the teeth of the notches. To maintain the reversing shaft rigid and so prevent this, Ivatt had applied his vacuum operated clutch in 1900 to that engine and the device was also adopted for 251.

Road trials showed where amendments to detail design of the engine were necessary. The large diameter of the smokebox required a chimney of reduced height. The blast-pipe top was 5 1/2" diameter and a petticoat extended downwards to terminate 10 5/8" above the blast-pipe nozzle, which was set on the centre line of the boiler tube-plate. When the final drafting work was being carried out, 251 carried a plain stovepipe chimney, which was replaced by one of a built-up pattern specially cast a few weeks later. Experiments were carried out with different chimney liner sizes and during April to September 1904, it ran with the liner extended well above the chimney top.

Although 251, was heavier than the 'Klondikes', the same arrangements of plate springs were retained for the trailing wheels. The riding left a lot to be desired, so the 3' 6" springs were replaced with 4' 6" springs, which had greater flexibility. A standard swing-link bogie was used and it was set back to the driving wheels. The axle-boxes were provided with helical springs but in September 1903, 251 was in the shops for a couple of days to have these supplemented by 'bogie control gear'. This mechanism consisted of small check springs which had a dampening effect on the springing; later engines had this arrangement when new.

Towards the end of 1903, four additional wash-out plugs, in staggered pairs of two, were located low down on each side of the firebox. Mounting of the brick arch was improved by notching the end bricks over the supports and the length of the arch was reduced by 9" by using smaller bricks. The front end of the firebox was then supported by a substantial steel casting which acted as a cross stretcher for the main frames while the fire-hole was shaped by dishing the inner and outer plates together and eliminating the ring. There was much attention to detail in that tweaking of different refinements was such that the whistle at the top front of the cab roof was brought an inch or two to the rear, so that the hole went through the curve of the backplate instead of through the wrapper.

As with many types of steam

The Pioneer large boiler Atlantic 251 emerged from Doncaster Works in December 1902, having cylinders, motion, wheel diameters and spacings similar to those on the preceding Klondikes, but with a variation to the main and auxiliary frames at the rear end. 251 ran with its original boiler until 24 January 1908. When it was withdrawn for preservation as the pioneer engine of its class in July 1947, it was restored to its 'as near as possible' original condition. It's seen here in preserved condition alongside a C12 class tank Atlantic engine in early BR lined black livery. The original circular lion and wheel or 'lion on a unicycle' logo – as seen on the tank side, appeared in about 1950 and replaced the words British Railways on tank and tender sides. This in turn was replaced by the rampant lion and wheel – otherwise known as the 'ferret and dart board' in 1956. (Author's Collection)

Built in June 1904, to works 1040, engine 283 was renumbered by the LNER twice and first became 3283 and then 2812. After it was withdrawn from service on 21 August 1947, it carried out stationary boiler duties at Doncaster Wagon Works from September 1947-June 1952.
(Author's Collection)

As smoke deflection was a problem in the case of Pioneer large boiler Atlantic N° 251, a wind deflector was fitted on the chimney cap in 1903, but was later removed and no further official changes were carried out. Then, for a short time in 1907, O.V.S. Bulleid arranged for a plain widely tapering chimney to be tried out on N° 251. Bulleid claimed that the engin steamed better, but when Ivatt saw this addition, he ordered that it be replaced by a standard one.
(Author's Collection)

engine design, smoke deflection was a problem with 251. So, in 1903, a wind deflector was fitted on the chimney cap but it was subsequently removed and after 1904, no further official chimney changes took place. However, a plain widely tapering chimney was tried out on 251, for a short period in 1907, at O.V.S. Bulleid's instigation. Bulleid at age 18 had joined the GNR at Doncaster as an apprentice under Ivatt the CME, and after a four-year apprenticeship he became the assistant to the Locomotive Running Superintendent, then a year later he was promoted to become the Doncaster Works manager. So based on draughting proportions evolved by Professor Goss, Bulleid claimed that the engine steamed perfectly and the exhaust was very much quieter. But when Ivatt saw this chimney, he made it clear to Bulleid that '... he was the Locomotive Engineer' by ordering '...and that it to be replaced by a standard one'. N° 251, ran with its original boiler until 24 January, 1908, but when it was withdrawn in July 1947, it was restored to original condition for preservation as the pioneer engine of its class.

With the provision of 32-element superheaters and adequate lubrication, the original 1902 design was developed to a high standard of performance. This probably reached its peak when all the standard Atlantics had received superheaters of the Robinson type during the early years of Grouping. The non-standard boiler of N° 1300 was fitted with a 22-element superheater in 1914 and this was retained, when, in 1917, Gresley converted it to a two-cylinder simple engine. It was scrapped in 1924 and was the first of the class to go.

N° 1421 also started out as a four-cylinder compound and appeared from the works in 1907, but didn't achieve a great measure of success and no further attempts at compounding were made. It was

This engine, 1421, started out as a four-cylinder compound design and appeared in this form when it was released from Doncaster Works in August 1907. However, the design didn't achieve a great measure of success and so no further attempts at compounding were made. It was then converted to a standard two-cylinder simple engine in 1921. It is seen in this image, in its compound design form, where it is seen in ex-works grey livery. (Author's Collection)

converted to a standard two-cylinder simple engine in 1921.

Alongside the new engine construction programme, Gresley instituted a series of improvements to the Atlantics. There was a constant effort to provide more efficient superheaters and new cylinders with piston valves were fitted to many of the standard engines. In 1915, Gresley rebuilt N° 279 as a four-cylinder engine with Walschaert's valve gear and rocking shafts to operate the valves for the inside cylinders, but reverted to a two-cylinder engine in 1928. It was given a contemporary appearance with a raised running plate, although the cab wasn't fitted with a side window.

The LNER was probably the only British railway company to experiment with the fitting of boosters, secondary steam engines provided on a steam engine's trailing axle or tender, used to assist with the starting of a train. The booster is intended to address two fundamental flaws of the standard steam engine. Firstly, most steam engines do not provide power to all wheels. The amount of force that can be applied to the rail depends on the weight resting on the driven wheels and the factor of adhesion of the wheels against the track. Unpowered wheels effectively 'waste' weight, which could be used for traction. Unpowered wheels are generally needed to provide stability at speed, but at low speed this is not necessary. Secondly, the gearing of a steam engine is constant, since the pistons are linked directly to the wheels via rods and cranks. Since this is fixed, a compromise must be struck between the ability to haul at low

Another view of 1421 when it was in compound design form. This image shows detail of the outside, high pressure cylinder design, which was almost identical to those of 292. 1421, was an improvement on 292 and the rearrangement of the low pressure cylinders resulted in a layout of pipe-work, which made it an easier matter to fit a superheater and so this was done in 1914. (Author's Collection)

Large boiler Atlantic 1421 began as a four-cylinder compound engine and appeared from Doncaster Works in 1907. It didn't achieve a great measure of success and was converted to a standard two-cylinder simple engine in 1921. It is seen in the rebuilt form here, working an express passenger service in GNR days. (Author's Collection)

speed and the ability to run fast without inducing excessive piston speeds, which would cause failure or the exhaustion of steam. This compromise means that the steam engine at low speeds was not able to use all the power that the boiler was capable of producing; it simply could not use steam that quickly and there was a big gap between the amount of steam the boiler could produce and the amount that could be used. The booster enabled the wasted steam potential to be put to good use. Therefore, to assist with starting a heavy train, some engines were provided with boosters to utilise this extra steam. Typically the booster engine was a small two cylinder steam engine back-gear-connected device connected to the trailing pony truck axle on the engine or, if none were available, the lead truck on the tender. A rocking idler gear permitted it to be put into operation by the driver. It would drive one axle only and could be non-reversible with one idler gear, or

Here is the GNR's 4-4-2 Atlantic 1410, built in August 1905, working a long and heavy express passenger service through Wood Green in 1923. (Author's Collection)

reversible with two idler gears. Used to start a heavy train or maintain low speed running under demanding conditions, the booster could normally be cut in while moving at speeds of under 15mph. Rated at about 300hp at speeds of 10 to 30mph, it would automatically cut out at 30mph.

Even before the formation of the LNER, Gresley considered the fitting of a booster to an Ivatt-designed C1 class Atlantic passenger express tender engine. So in October 1922, he ordered an American built booster. This was a two cylinder steam engine which was adapted to drive the trailing wheels of a steam engine and was to be used when extra power was needed at starting or climbing a steep gradient.

It is often said that the camera doesn't lie, but here is tangible proof that as loadings increased, the available motive power wasn't up to the job. This picture shows that on this particular working the motive power was 'beefed-up', with the hauling power of a single engine deemed to be insufficient to do the job effectively, with the more powerful of the two engines, the large boiler Atlantic being insufficient to easily work this very long and heavy service. In this image, Ivatt class D1 class 4-4-0 engine 65 is being piloted by Ivatt C1 class 4-4-2 large boiler Atlantic 1430. The D1s were Ivatt's final 4-4-0 class and were built in 1911. The first five D1s were withdrawn in 1946, with the last D1 being withdrawn in November 1950. 1430 was completed in April 1907 and was withdrawn on 3 February 1944. (Author's Collection)

C1 class Atlantic 4443, seen here, was renumbered a second time by the LNER to become 2873. (Author's Collection)

The LNER was the only British railway company to experiment with the fitting of boosters, which consisted of a two cylinder steam engine adapted to drive the trailing wheels of a steam engine, to provide extra power when needed at starting or when climbing a steep gradient. Although there was a design of steam tender which had its own engines that was introduced in 1863, by Archibald Sturrock on the Great Northern Railway. Before the formation of the LNER, Gresley, as the locomotive superintendent of the GNR, considered the fitting of a booster to an Ivatt-built C1 class Atlantic engine. So in October 1922, he ordered an American built booster and 1419 was fitted with this booster to the trailing wheels in October 1923. The image shows what a booster looked like. (Author's Collection)

The Ivatt-designed C1 class Atlantic engine N° 1419 acquired this booster, which was fitted to its trailing wheels in October 1923 and it was renumbered by the LNER and became N° 4419. The booster, in theory, improved the tractive effort by about 50 per cent and the booster could be engaged by means of compressed air, which was supplied by a Westinghouse pump. It had cylinders that were 10" x 12" and they were connected to the axle with gears of total ratio 36:14. After extensive trials around the Bishop Auckland area and with many refinements, including a new gear ratio of 36:24, Gresley was still not too happy about the riding qualities of 4419 and wrote a letter that was to lead the way to another class of steam engine being fitted with boosters. Because of these riding problems and other mechanical difficulties, the booster was subsequently removed from 4419 in 1935.

However, results were encouraging enough for Gresley to have boosters fitted to his two new large freight engines, the mineral P1 class 2-8-2 Mikado engines. The first of this pair, N° 2393, appeared in June, 1925, with the second engine, N° 2394, appearing in November of the same year. Despite being well liked, early trials showed that the boosters were only effective with loads of at least 1,600 tons, and because typical loads were much less than this, and with the additional problem of leakage in the steam pipes leading to the boosters, the boosters were removed in 1937 on 2394 and in 1938 from 2393.

In the summer of 1923, C1 class Atlantic N° 1447 received cut down

When large boiler Atlantic N° 4419 was booster fitted, it had extensive trials around the Bishop Auckland area. The result was that it had many refinements, including the fitting of a new gear ratio of 36:24. But the riding qualities of N° 4419 were considered by Gresley to be unacceptable and so the booster was removed in 1935. Here N° 4419 is seen at King's Cross in 1927 (Author's Collection)

Because of the poor riding qualities of 4419, another class of engine was fitted with boosters instead, namely the P1 class 2-8-2 freight engines. There were only two members of the 2-8-2 P1 class and they were two of the most powerful freight engines ever built for a British railway, with the power being quoted as being 25 per cent more than the O2 class. The image shows 2393, the first member of the class. (Author's Collection)

boiler mountings and the cab profile was altered to follow the contours of the North British loading gauge. This was to allow the engine to take part in competitive trials held that year, with Atlantics representing the former NER and NBR. Thoughts turned in 1936, to a possible scheme to streamline one or two members of the class, but this didn't happen.

All brakes on the C1 class Atlantics had the standard vacuum brake operating on the coupled wheels. The brake gear on the latter however, were removed from N° 1419, when it was later fitted with a booster. The standard engines had two 18' diameter brake cylinders fitted side by side between the frames behind the bogie, actuating the brakes on the coupled wheels. A third cylinder, also 18" diameter, was located under the cab for the trailing wheels. N° 279, differed from the standard arrangement after rebuilding in 1915, by having 21" diameter cylinders fitted behind the bogie instead of 18' ones.

The two compound engines, 1421 and 292, had brakes that differed from the standard arrangement by having their 18' diameter brake cylinders for the coupled wheels positioned behind the leading coupled wheels for clearance reasons. N° 1421, reverted to the normal arrangement when it was rebuilt in 1920 as a standard Atlantic. N° 1300 differed from the other engines both before and after its 1917 rebuilding, by having its brake cylinders all located under the cab. The brake blocks, therefore, acted against the front of the coupled wheels instead of the rear, as in all other engines in the class. The vacuum ejector exhaust pipe passed through the boiler and in contrast to many other classes it was never altered to run along the outside.

Class B, well-bottom type six-wheel tenders on an equally divided wheel-base of 13' which were fitted with water-scoop pick-up apparatus, were attached to the standard large boiler Atlantics, being built up to 1907. The tenders had a water capacity of 3,670 gallons, with a coal capacity of 5 tons, giving a weight when full of 40 tons 18cwt. The water-scoop was operated by a vertical lever which was located to the left of the coal gate and the water-tank filler and pick-up dome, which were combined in a single rectangular casing, with a hinged lid at the rear. Two coal rails were provided to prevent spillage. They were originally open but were, from 1908, fitted with backing plates for the length of the coal space. When new in 1905, N° 1300 had a similar class B tender, but it was not fitted with water-scoop apparatus. A number of engines were observed running attached to Stirling tenders, which were easily recognisable by their wooden buffer beams.

In 1908, a new pattern of six-wheel tenders was introduced, which had a reduced water capacity of 3,500 gallons and an increased coal capacity of $6^1/_2$ tons, giving an increase in weight, fully laden, from

Large boiler Atlantic 276 had side chains fitted to the buffer beams when it was new, with the figuring set high on the beams. After the removal of the chains in 1905, the numbers were centralised and this style was also applied to all subsequent new engines. Commencing about 1909, the size of the numerals was increased to 6" but Gresley reverted to $4^1/_2$" shaded characters which then became standard. 276, is seen leaving King's Cross Station with its side chains removed and the numerals centralised. (Author's Collection)

40 to 43 tons 2cwt. In the new design, the wheelbase was this time unequally divided, with the distance of the wheels from the front set at 6' and 7' distance. This time the brake blocks acted in the usual way on the rear, instead of on the front of the wheels, as was the case in the preceding 3,670 gallon design tenders.

Atlantics constructed from 1908 had the new design tenders attached to them when they were built, with the earlier built engines receiving them as replacements. By the early 1920s, all large boiler Atlantics were equipped with 3,500 gallon tenders, although there were some external detail differences. The 1908 design, had cut-away corners to the side sheets at the front end, to which curved handrails were added later. The tenders that were provided as replacements, had plain rounded corners with no cut-away. The vertical scoop lever was retained as standard, but the design of the coal gate and location of the stowage lockers, which replaced the separate toolboxes bolted to the tank tops in the pre-1908 tenders, varied between the earlier and later batches. A separate water pick-up dome and filler were provided and both were circular in plan. The coal rails were backed by $^1/_4$" steel plate.

After rebuilding in 1917, a new standard class B tender, fitted with water-scoop pick-up apparatus, was attached to N° 1300. Several tenders had three coal rails instead of the usual two and these ran with various Atlantics over a period of time, for example on numbers 279, 284, 295, 1411, 1427, 1458 and 1460. It was common practice to interchange tenders and other engines acquired such tenders after the Grouping.

Between May and November 1921, six engines, 277, 289, 1402, 1432, 1439 and 1448, were repaired at Beardmore's, Glasgow, as part of a Government scheme to provide work at factories, which had lost work after the First World War. The total cost of this contract was £17,016. Prior to leaving Doncaster for Glasgow, the springs on these engines were adjusted to lower their overall height to 13' 2", due to the stricter loading gauge that was in force in the Scottish region.

The livery of the large boiler Atlantics was standard apple green, lined out in black and edged with a small white line and all wheel centres were green. The smokebox and cylinder covers were plain black. On N° 251 only the bottom edge of the firebox cladding was originally finished with a black panel some 6' deep. The engine frames, footsteps, motion brackets and valances were dark red -brown, with fine vermilion and black lining. Buffer beams were vermilion with an edging of black, which, circa 1910, was changed to a black and white border. The engine running numbers were carried on the front buffer beam in $4^1/_2$" gilt numerals, which were blocked, in black and brown to the right-hand side.

As a result of Atlantics numbers 251, 273-92 and 1300 having side chains fitted to the buffer beams when new, the figuring was set high on the beams. After the removal of the chains in 1905, the numbers were centralised and this style was also applied to all subsequent new engines. Commencing about 1909, the size of the numerals was increased to 6" but Gresley reverted to 4¹/₂" shaded characters, which then became standard. The reverse-curved lining of the panelling on the sides and rear of the tender left surrounds were painted in holly green. Lettering on the tender sides was 7¹/₂" high, in gold-leaf, blocked in red and black, to the right-hand edge.

When new, 251 and 273-92 had small cast oval Doncaster Works plates, which had raised lettering with red backgrounds, bolted to the main frames immediately to the rear of the smokebox. On 251, 272, 274-5 and 278, these plates were later moved to the frames forward of the smokebox, as were those of 279 when it was rebuilt in 1915. N° 251's works plates were returned to their original location when the engine was restored for preservation in 1947.

Compound C1 class engine N° 292 appeared in 1905, with plates of a similar pattern fitted in the usual position to the rear of the smokebox, but they were soon removed when the addition of sheet metal casings were provided to protect the valve rods above the running plate level. New, larger oval plain brass works plates, with incised lettering were then introduced and they were mounted in the centre of the leading splashers. This style and location was thereafter adopted for all subsequent Atlantics as built. N° 1300 had Vulcan Works plates fitted on the leading couple-wheel splashers, with the same location used for the replacement ones after its rebuilding. Both types of plates were small oval brass ones, with raised lettering on a red background.

When large boiler Atlantic 301, seen in the image, was withdrawn from service on 22 July 1948, it carried out stationary boiler duties at Doncaster Carriage Works from July 1948 to August 1953. It was finally cut up at Doncaster in August 1953. (Peter N. Townend Collection)

Towards the end of December 1922, early consideration was given to the selection of liveries pending the 1923 Grouping, whereby N° 286 and N° 1429 appeared ex-Works in GN livery, but they were devoid of their company initials on the tender sides. Early in 1923, N° 1418 was painted in GN two-tone green, with reverse curves on the tender panelling but it was re-lettered to represent GN passenger classes, in an array of engines drawn up for inspection by a Board committee at York old Station on 31 January 1923. Prior to this exhibition, Doncaster had tried out an alternative layout on the left-hand side only, of the tender attached to 1418, where the new company initials LNER appeared in an arc, with the running number located beneath them and on the cab sides. Gresley disliked this style and this version was painted out the same day. Following a second exhibition, this

Built at Doncaster Works in June 1905, LNER C1 class 4-4-2 engine No 4402 is seen to effortlessly work its express pasenger service. (Author's Collection)

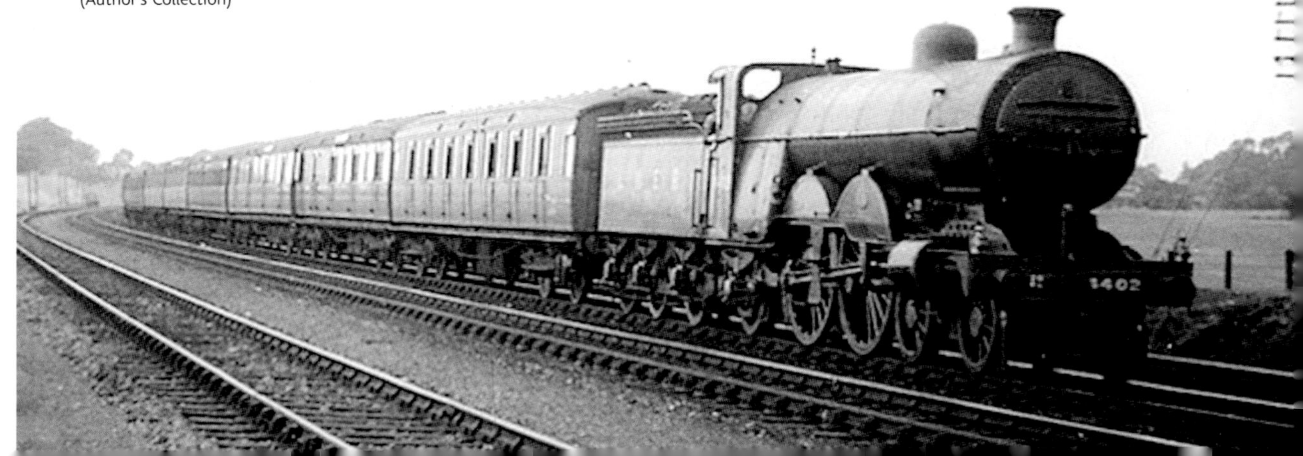

On 19 September 1906, the train being worked by 276 took a crossover at too great a speed, resulting in the engine turning over on its left side with the tender separating and falling down the embankment. The driver and fireman were both killed. This image is of 276, after the LNER had renumbered it to become 3276. (Author's Collection)

time held on 23 February at Marylebone Station, the committee decided upon GN green for all engines having 6' 6" diameter wheels or over.

When the details were resolved, the new livery differed from the old version by the deletion of the two-tone green panels from the tender, which had plain curved lettering and the substitution of black for red-brown parts on the engine, which was given a 3/16" vermillion line, $^1/_2$" from the edges. The new company initials later omitted the ampersand and the preferred version as agreed was LNER instead.

Few large boiler Atlantics were involved in serious accidents, but the Grantham derailment of 19 September 1906 became famous when a sleeper train worked by N° 276 took a crossover at excessive speed through a station after passing a signal at red. The engine turned over on its left side with the tender breaking loose and falling down the embankment. Driver Fleetwood of Doncaster and Fireman Talbot, who had been a premium apprentice at Doncaster Works for five years were both killed. A total of fourteen people were killed and seventeen were injured. On 13 February 1923, as 298 worked an express passenger train it too overran signals and was in a rear-end collision with a freight train at Retford, Nottinghamshire. In this incident three people were killed. On 15 June 1935, 4411 was involved in an incident whereby as it was working a passenger service, the train was run into by an express passenger train at Welwyn Garden City, due to a signalman's error. Fourteen people were killed and twenty-nine were injured

Most of the work undertaken by Pioneer large Atlantic N° 251 during the eighteen month period between its completion in December 1902 and the arrival of the first production batch, was between Peterborough and Doncaster. In 1903, revised engine working diagrams introduced through running each way between Doncaster and London. On 17 September 1903, 251 undertook its first departure out of King's Cross and by October it was regularly working into King's Cross with the 10:40 Mark Lane Express and departing at 13:30 with the Leeds - Bradford Express on alternate days. The opportunity was taken to have a test run with 251, using the GWR's Dynamometer Car, working the 14:20 service. The Dynamometer Car was on its way to the NER, who were going to use it on test runs with their V class Atlantic engine. No official details appear to have been released concerning the performance during this test run.

As the large boiler Atlantics took to the road, their drivers were presented with an engine far bigger than that to which they had previously been accustomed. The allocation of one engine to a particular driver enabled him to acquaint himself with the best possible practice for operation and maintenance. From the earliest days of operation of the large boiler Atlantics, it was reported by some that the firemen wore unofficial leather aprons to protect their legs from the heat from the firebox

1442 was chosen to become the 'Royal' engine and so in keeping with the current arrangements it was specially decorated with the GNR coat of arms on its rear splashers. It's seen here with its six coach Royal Train. (Author's Collection)

whilst they were firing.

By November 1905, forty-three large boiler Atlantics were at work with seven engines allocated to London, nine allocated to Peterborough, thirteen at Grantham and fourteen examples at Doncaster, including the Compound Atlantics 292 and 1300. By 1906, comparative trials were conducted with these two and the standard Atlantic 294 which was working at 200lb psi boiler pressure. The trains worked were between King's Cross and Doncaster. All engines worked similar trains and used the same type of coal; Yorkshire, from the Barnsley bed. An inspector rode with each engine during the trials and careful account was taken of coal and oil used, time gained or lost etc and the cost of running repairs. The chart on pages 108-109 shows the relevant figures obtained.

When N° 1442 was chosen to become the 'Royal' engine, in keeping with the current arrangements, it was specially decorated with the GNR coat of arms on the rear splashers. Its wheel rims, buffers, brake, stand pipe and other exposed metal parts were burnished bright, with polished brass rims to the coupled-wheel splashers. It retained the crests and brass-rimmed splashers until it was withdrawn from service on 23 January 1947.

When 1442 was chosen to become the 'Royal' engine, it was specially decorated with the GNR coat of arms on the rear splashers as seen here. 1442 retained the crests and brass-rimmed splashers until it was withdrawn from service on 23 January 1947. (Peter N. Townend Collection)

It made its first trip with the Royal train on 7 July 1908, conveying King Edward VII from King's Cross to Leeds. The engine was taken out of traffic early in 1909, after a mileage of 48,000 to be prepared to a high finish for exhibition in the Machinery Hall at the White City, Shepherds Bush, London, together with Patrick Stirling's 4-2-2 '8-footer' N° 1, of 1870. To get the engines to Shepherds Bush a special train was formed. It was worked by a rebuilt Stirling 2-4-0 N° 867 and consisted of 1442 and N° 1, which were both sheeted over and with special buffer padding, together with five wagons of permanent way material, one tool van and a bogie brake. The ensemble departed from Doncaster at 09:00 on Friday, 23 April 1909 bound for London. With cautious running and examination stops made at Retford, Newark, Grantham, Peterborough and Hitchin, the train finally arrived at Hornsey at 20:20 later that day, where it stayed until Sunday, 25 April. The 0-6-0 saddle tank N° 1201, then took the train forward to Finsbury Park, Canonbury and Dalston to Kingsland where a North London engine took over for the next stage of the journey via Hampstead Heath and Willesden to Acton. The GWR then worked the formation to Uxbridge Road goods yard, which adjoined the grounds laid out for the Imperial International Exhibition.

When the engines were on show, they stood on permanent way of their respective periods. The 1870 rails were of 80lb per yd, compared with 100lb per yd for 1909, which also had a section of water trough laid out on the foreground.

After the exhibition closed on 25 September 1909, the same means

Large boiler Atlantic 1442 is seen exhibited at the Imperial International Exhibition at White City, London. Alongside is another GNR engine, none other than Stirling's 'Eight Footer' 4-2-2 N° 1 (Author's Collection)

Here, the L&NWR 'Precursor' class 4-4-0 N° 412, Marquis, is seen with a train of GNR carriages on the GNR main line during interchange trials during 1909. (L&NWR Society Collection ~ N° NHL 17)

was employed on 12 December to return both engines to King's Cross, after which 1442 returned to service and N° 1, was put into store there. Interestingly, they say that history has a way of repeating itself and it did on this occasion, not only once but twice, some 101 years later. Stirling Single N° 1 went 'Mainline' again on 2 June 2010, when it was moved from Southall depot in west London to Waterloo International via the West London Line, retracing its route and passing the former White City Exhibition site. It was being moved to Waterloo to appear as the star player in the stage adaptation of E. Nesbitt's *The Railway Children*, staged on the site of the former International Station's platforms. The train's formation for the move consisted of: N° 37516, which was at the London end; diesel shunters N° 08943 and N° 08911 (named *Matey*) which would move the 'old gentleman's' train into and out of the show at Waterloo; two mark one coaches; 'The Old Gentleman's Saloon' N° 21661, which was made famous in the film *The Railway Children*; N° 1 and N° 37706, at the country end. After a successful run, the show closed and on 5 January 2011, the return journey from Waterloo International to Southall was made via the West London Line, but this time the train was topped and tailed by two class 47 diesel engines, with N° 47500 at the London end and N° 47760 at the Southall end.

Locomotive exchange trials were held for one month commencing on Thursday, 10 July 1909, in conjunction with the LNWR, who used its Precursor class against the GNR's large boiler Atlantic. On the GNR, the LNWR's 4-4-0 N° 412 Marquis was duly sent to Doncaster for this purpose and was coupled to a standard Atlantic tender painted black. Doncaster-based large boiler Atlantic N° 1451, ran alternatively as the competing engine each of which covered a weekly total of 2,050 miles accumulating mileage on various trips between London, Doncaster and Leeds. Houghton coal was consumed on the trials and an economy of 6 per cent per ton-mile was shown in favour of Atlantic 1451.

For the corresponding trials over the LNWR main line, N° 1449 was shedded at Camden. The competing engine was N° 510, *Albatross*, with twenty-six trips being made by both engines over the same routes from Euston to Crewe, on alternate days, covering 4,113 miles with an average load of 320 tons. The conclusion of the trials was that 1449 showed an 8.5 per cent advantage in coal economy over the LNWR's 4-4-0 aptly named *Albatross*.

From 1910 until 1922, the large boiler Atlantics dominated main line express duties, helped out at times by their smaller brothers the 'Klondikes'. Prior to superheating, the performance of the large Atlantics could be erratic. They were slow starters, particularly

There's snow everywhere as the GNR's large boiler Atlantic 1449 waits at Euston Station, London, with a long up-express of L&NWR carriages, during the 1909 interchange trials. 1449 was built in August 1908 at Doncaster Works and was withdrawn in July 1947 from Grantham MPD. (Author's Collection)

when loads were heavy. Until 1908, there were no water troughs south of Werrington, which is eighty miles out from King's Cross, so maybe the drivers were holding back for fear of using too much water, too soon. Then in 1922, Gresley's A1 class 4-6-2 Pacifics started emerging from Doncaster Works and so moved the Atlantics onto less important duties, although they lasted until Sunday 26 November 1950. On this day forty-five years or so after first emerging from Doncaster Works, the last working Atlantic on BR's Eastern Region, N° 62822 ended its days in a blaze of glory by departing from King's Cross Station at 11:00, bound for Doncaster. It was an occasion tinged with sadness for all of those who had been associated with the Ivatt Atlantics. However, the Pioneer of the class N° 251, has been preserved.

In the 1900 engine classification scheme, both 'Klondikes' and large Atlantics were classified as C1 class. There were no sub-divisions to distinguish boiler size, compounds etc. Under the LNER, the large boiler Atlantics were designated as C1 class, including the three experimental engines numbered 279, 292 and 1300. From 1924, all of the C1s, with the exception of the Vulcan 1300, had 3000 added to their running numbers. Under the comprehensive renumbering scheme prepared by Edward Thompson in 1943, the surviving engines were allocated the numbers 2800-91 in running number order. When the scheme had been published, 4459 had been withdrawn and its new number (2889) had been left blank. Heavy withdrawals had depleted the number of engines by January 1946 when the scheme was implemented and only fifty Atlantics later received their new numbers.

A4 class Pacific N° 4468, *Mallard*, is the holder of the world speed

GNR large boiler 4-4-2 1449, accelerates away from Crewe in a southerly direction, with a very long up-express of L&NWR carriages, during the 1909 interchange trials. On the extreme left are the Crewe South carriage sheds and in front of them are several L&NWR signals, some of which are ringed and are located on tall posts. (L&NWR Society Collection ~ N° SOC 1236)

This is the scene at Sheffield on 2 June 1924, of the arrival of the very first 'Sheffield Pullman' service, worked from King's Cross by large boiler Atlantic 4426. (Antony M. Ford Collection)

The very first down working of the 'West Riding Pullman' was worked by A1 class Pacific 4474, *Victor Wild*, on 15 May 1930. Although there were two fewer Pullman Cars in its formation, here large boiler Atlantic 4454 is seen working this prestigious train. The reduction in load was due to the Atlantic engine's reduced power, as compared with that of a Pacific engine. (Authors Collection)

record for steam engines at 125.88mph. The record was achieved on 3 July 1938 on the slight downward grade of Stoke Bank south of Grantham on the East Coast Main Line and the highest speed was recorded at milepost 90¼, between Little Bytham and Essendine. It broke the German DRG class, N° 05002's 1936 record of 124.5mph. The record attempt was carried out during the trials of a new quick acting brake, the Westinghouse 'QSA' brake. On the record run Mallard arrived at Barkston at 14:49, where the train and engine were turned and parked into a siding. A late lunch was taken, washed down with cups of tea. During the layover, Sam Jenkins and Eric Bannister thoroughly doused the middle big end with super-heater oil. After lunch Jenkins and Tommy Bray made up the fire and re-tested the intercom equipment in readiness for the high speed run. The fire was made as hot as possible to generate the steam required. A quick roll call was done. Just before departure Douglass Edge called the Westinghouse team together. He addressed a M. La Clair telling him what was about to happen and asked if anyone wanted to take a taxi back to Peterborough; needless to say everyone stayed. *Mallard* gave a couple of short whistles and slowly ambled out onto the mainline. The time was 04:15 precisely. *Mallard* was checked at Grantham nearly 'wrecking the run', a PW (permanent way) slack reduced the speed to 18mph so she had to be coaxed to regain speed. *Mallard* passed Stoke Signal Box at

This is the scene as a 'rag-bag' of different coaches as was used for a LNER main line express service, worked by large boiler Atlantic 3278, leading a D1 class 4-4-0 engine. (Author's Collection)

LNER C1 class Atlantic 4429, leads a mixed train of four-wheel and bogied stock during LNER days. (Author's Collection)

85mph and the rest, as they say, is history.

Steam was shut off for Essendine's 90mph curves and at this point the middle big end overheated and melted the white metal. *Mallard* rounded the curves at Essendine at 108mph and was reported to have leaned over most alarmingly and there was nearly the fastest high speed steam wreck, but luck held. Fitters swarmed over the engine when it came to rest at Peterborough North after having limped in. It was, of course, unable to run through to London and was quickly replaced by an Ivatt large boiler Atlantic, N° 3290, which had originally entered service in June 1904. It took the train onwards to Platform 1, King's Cross and arrived at exactly 18:27, but, of course, minus *Mallard*. The Press, who were waiting at King's Cross to meet the heroes of the high-speed run and the train, were invited into the Dynamometer Car to inspect the record charts, while N° 3290 was spirited away to King's Cross's Top-Shed. The record breaking high-speed run was over!

Gresley wasn't totally satisfied with the run and planned more, but war stopped them happening. He felt that *Mallard* was good for 130mph, pushing onto even 140mph. Had the PW check not slowed the engine then perhaps the record would have been higher and *Mallard* would have slowed down more gently and sooner, thus alleviating the melted middle big-end white metal bearing. Maybe – we will never know!

History of GNR 4-4-2 Large 'ATLANTIC' locomotives ~ 1902 - 1950

GNR Number	Date to Traffic	Doncaster Works Number	First LNER Number ~ () = not carried	Second LNER Number ~ () = not carried	BR Number () = not carried	Fitted with Superheater	Converted with Piston Valves	Rebuilt to 4-cylinder compound engine	Rebuilt to a 2-cylinder simple engine	Fitted with booster	Shed Allocation in June 1930	Withdrawal Date	Scrap Date	Where Cut Up
Nº 251	Completed in December 1902 - to traffic January 1903	Nº 991	Nº 3251	Nº 2800	-	August 1918	December 1923	-	-	-	King's Cross	28th July 1947	-	Preserved at the NRM, York
Nº 272	May 1904	Nº 1030	Nº 3272	Nº 2801	-	February 1916	February 1916	-	-	-	-	8th August 1947	-	Doncaster
Nº 273	June 1904	Nº 1031	Nº 3273	Nº 2802	-	March 1916	March 1916	-	-	-	-	31st January 1947	-	Doncaster
Nº 274	June 1904	Nº 1032	Nº 3274	(Nº 2803)	-	June 1920	June 1920	-	-	-	King's Cross	2nd May 1946	March 1952	Carried out Stationary Boiler duties at Don. Loco Works, from September 1947, finally cut in March 1952
Nº 275	June 1904	Nº 1033	Nº 3275	(Nº 2804)	-	August 1921	Slide Valves until w/drawn	-	-	-	-	5th May 1946	-	Doncaster
Nº 276	June 1904	Nº 1034	Nº 3276	(Nº 2805)	-	October 1918	Slide Valves until w/drawn	-	-	-	-	1st January 1946	-	Doncaster
Nº 277	June 1904	Nº 1035	Nº 3277	(Nº 2806)	-	April 1919	Slide Valves until w/drawn	-	-	-	-	31st January 1945	-	Doncaster
Nº 278	June 1904	Nº 1036	Nº 3278	(Nº 2807)	-	September 1920	September 1920	-	-	-	King's Cross	26th July 1945	1988/89	Nº 3278's boiler, was used at Maldon Sawmill circa 1948 until 1975. Boiler scrapped at Brightlingsea circa 1988-89.
Nº 279	June 1904	Nº 1037	Nº 3279	Nº 2808	(Nº 62808)	May 1915	May 1915	-	1938	-	King's Cross	14th February 1948	March 1948	Doncaster
Nº 280	June 1904	Nº 1038	Nº 3280	(Nº 2809)	-	July 1922	January 1931	-	-	-	-	26th October 1946	-	Doncaster
Nº 281	June 1904	Nº 1039	Nº 3281	Nº 2810	(Nº 62810)	July 1920	October 1935	-	-	-	-	27th May 1949	July 1949	In for cut up @ Doncaster 1st June 1949
Nº 282	June 1904	Nº 1042	Nº 3282	Nº 2811	-	August 1917	Slide Valves until w/drawn	-	-	-	-	30th January 1947	-	Doncaster
Nº 283	June 1904	Nº 1040	Nº 3283	Nº 2812	-	August 1923	December 1924	-	-	-	-	21st August 1947	-	Carried out Stationary Boiler duties at Doncaster Wagon Works from September 1947 - June 1952
Nº 284	June 1904	Nº 1044	Nº 3284	(Nº 2813)	-	September 1917	September 1917	-	-	-	King's Cross	26th January 1946		Doncaster
Nº 285	June 1904	Nº 1041	Nº 3285	(Nº 2814)	-	March 1918	December 1919	-	-	-	-	2nd May 1946	-	Carried out Stationary Boiler duties at Don. Loco Works, from September 1947 until March 1952
Nº 286	June 1904	Nº 1043	Nº 3286	Nº 2815	-	May 1924	Slide Valves until w/drawn	-	-	-	King's Cross	17th October 1947	-	Doncaster
Nº 287	June 1904	Nº 1045	Nº 3287	(Nº 2816)	-	June 1922	Slide Valves until w/drawn	-	-	-	-	6th October 1945	-	Nº 3287's boiler Nº 9359 - new in August 1943, was moved with Nº 3278's boiler - see above, from Doncaster to Maldon Saw Mill in the late-1940's. Now being restored at the Bluebell Railway
Nº 288	June 1904	Nº 1047	Nº 3288	Nº 2817	(Nº 62817)	March 1918	December 1919	-	-	-	King's Cross	8th May 1950	August 1950	Doncaster
Nº 289	June 1904	Nº 1046	Nº 3289	(Nº 2818)	-	February 1919	February 1919	-	-	-	-	20th December 1945	-	Doncaster
Nº 290	June 1904	Nº 1048	Nº 3290	(Nº 2819)	-	April 1920	June 1922	-	-	-	-	9th November 1945	-	Doncaster
Nº 291	June 1904	Nº 1049	Nº 3291	(Nº 2820)	-	July 1922	Slide Valves until w/drawn	-	-	-	-	21st April 1945	-	Doncaster
Nº 292	March 1905	Nº 1066	Nº 3292	-	-	-	-	1905	-	-	-	5th January 1927	August 1928	Doncaster
Nº 293	April 1905	Nº 1067	Nº 3293	Nº 2821	(Nº 62821)	October 1919	July 1926	-	-	-	-	2nd July 1948	August 1950	Carried out Stationary Boiler duties at Don. Carriage Works from July 1948 to August 1950. Cut up @ Don.
Nº 294	May 1905	Nº 1070	Nº 3294	Nº 2822	Nº 62822	September 1917	September 1917	-	-	-	-	27th November 1950	May 1951	Doncaster
Nº 295	May 1905	Nº 1071	Nº 3295	Nº 2823	-	October 1915	October 1915	-	-	-	King's Cross	8th August 1947		Doncaster
Nº 296	April 1905	Nº 1068	Nº 3296	(Nº 2824)	-	August 1917	Slide Valves until w/drawn	-	-	-	-	7th December 1945	-	Carried out Stationary Boiler duties at Don. Carriage Works as s/b Nº 798 from December 1945 to circa 1965.
Nº 297	April 1905	Nº 1069	Nº 3297	Nº 2825	-	September 1923	August 1932	-	-	-	-	5th February 1947	-	Doncaster
Nº 298	May 1905	Nº 1072	Nº 3298	(Nº 2826)	-	February 1917	February 1917	-	-	-	-	4th April 1945	-	Doncaster
Nº 299	May 1905	Nº 1073	Nº 3299	(Nº 2827)	-	June 1919	Slide Valves until w/drawn	-	-	-	King's Cross	25th May 1945	-	Doncaster
Nº 300	June 1905	Nº 1075	Nº 3300	Nº 2828	(Nº 62828)	January 1918	January 1920	-	-	-	King's Cross	29th September 1949	September 1949	In for cut up @ Doncaster 30th August 1949
Nº 301	May 1905	Nº 1074	Nº 3301	Nº 2829	(Nº 62829)	March 1917	March 1917	-	-	-	King's Cross	22nd July 1948	August 1953	Carried out Stationary Boiler duties at Don. Carriage Works from July 1948 to August 1953. Cut up @ Don.
Nº 1300	July 1905	Nº 2025 - Vulcan Works number	(Nº 4300)	-	-	November 1914	November 1917	1905	1917	-	-	25th October 1924	-	Doncaster
Nº 1400	June 1905	Nº 1076	Nº 4400	Nº 2830	-	June 1921	Slide Valves until w/drawn	-	-	-	-	6th December 1947	-	Doncaster
Nº 1401	June 1905	Nº 1077	Nº 4401	Nº 2831	-	September 1915	September 1915	-	-	-	-	6th November 1946	-	Doncaster
Nº 1402	June 1905	Nº 1078	Nº 4402	Nº 2832	-	August 1917	February 1934	-	-	-	-	5th August 1947	-	Doncaster
Nº 1403	June 1905	Nº 1079	Nº 4403	Nº 2833	-	August 1916	August 1916	-	-	-	-	7th July 1947	-	Doncaster
Nº 1404	June 1905	Nº 1081	Nº 4404	Nº 2834	-	March 1914	March 1914	-	-	-	-	8th August 1947	-	Darlington
Nº 1405	June 1905	Nº 1080	Nº 4405	Nº 2835	-	April 1914	April 1914	-	-	-	-	14th August 1947	-	Darlington
Nº 1406	June 1905	Nº 1082	Nº 4406	(Nº 2836)	-	April 1914	April 1914	-	-	-	-	16th February 1946	-	Doncaster
Nº 1407	July 1905	Nº 1083	Nº 4407	(Nº 2837)	-	August 1915	August 1915	-	-	-	-	6th June 1944	-	Doncaster
Nº 1408	July 1905	Nº 1084	Nº 4408	(Nº 2838)	-	November 1921	Slide Valves until w/drawn	-	-	-	-	26th July 1945	-	Doncaster
Nº 1409	July 1905	Nº 1085	Nº 4409	Nº 2839	(Nº 62839)	April 1917	April 1917	-	-	-	-	23rd January 1950	January 1950	In for cut up @ Doncaster 26th January 1950
Nº 1410	August 1905	Nº 1086	Nº 4410	Nº 2840	-	July 1917	August 1919	-	-	-	-	29th April 1947	-	Doncaster
Nº 1411	March 1906	Nº 1109	Nº 4411	Nº 2841	-	May 1924	November 1926	-	-	-	King's Cross	6th August 1947	-	Darlington
Nº 1412	March 1906	Nº 1110	Nº 4412	Nº 2842	-	June 1916	June 1916	-	-	-	-	20th May 1947	-	Doncaster
Nº 1413	March 1906	Nº 1111	Nº 4413	(Nº 2843)	-	September 1923	September 1923	-	-	-	-	4th January 1946	-	Carried out Stationary Boiler duties at Don. Carriage Works as s/b Nº 810 from January 1947 to circa 1965.
Nº 1414	March 1906	Nº 1112	Nº 4414	(Nº 2844)	-	June 1918	Slide Valves until w/drawn	-	-	-	-	28th October 1944	-	Doncaster
Nº 1415	April 1906	Nº 1113	Nº 4415	Nº 2845	-	June 1920	April 1927	-	-	-	-	29th April 1947	-	Darlington
Nº 1416	April 1906	Nº 1114	Nº 4416	(Nº 2846)	-	January 1915	April 1915	-	-	-	-	20th October 1945	-	Doncaster
Nº 1417	April 1906	Nº 1115	Nº 4417	(Nº 2847)	-	May 1947	May 1917	-	-	-	-	21st April 1945	-	Carried out Stationary Boiler duties at Don. Carriage Works as s/b Nº 810 from January 1947 to circa 1965.

GNR Number	Date to Traffic	Doncaster Works Number	First LNER Number ~ ( ) = not carried	Second LNER Number ~ ( ) = not carried	BR Number ( ) = not carried	Fitted with Superheater	Converted with Piston Valves	Rebuilt to 4-cylinder compound engine	Rebuilt to a 2-cylinder simple engine	Fitted with booster	Shed Allocation in June 1930	Withdrawal Date	Scrap Date	Where Cut Up
Nº 1418	May 1906	Nº 1116	Nº 4418	(Nº 2848)	-	June 1917	August 1920	-	-	-	-	20th February 1945	-	Doncaster
Nº 1419	May 1906	Nº 1117	Nº 4419	Nº 2849	(Nº 62849)	July 1923	July 1923	-	-	1923 - 1935	King's Cross	19th July 1948	July 1953	Carried out Stationary Boiler duties at Don. Carriage Works from July 1948 to July 1953. Cut up @ Don.
Nº 1420	May 1906	Nº 1118	Nº 4420	Nº 2850	-	May 1919	August 1928	-	-	-	King's Cross	5th August 1947	-	Doncaster
Nº 1421	August 1907	Nº 1166	Nº 4421	Nº 2851	-	December 1914	December 1920	1907	1920	-		5th August 1947	-	Doncaster
Nº 1422	March 1907	Nº 1144	Nº 4422	(Nº 2852)	-	June 1919	October 1924	-	-	-	King's Cross	28th September 1946	-	Doncaster
Nº 1423	April 1907	Nº 1146	Nº 4423	Nº 2853	-	August 1919	August 1919	-	-	-		27th May 1947	-	Doncaster
Nº 1424	April 1907	Nº 1147	Nº 4424	Nº 2854	(Nº 62854)	April 1924	September 1935	-	-	-	King's Cross	3rd April 1950	May 1950	Doncaster
Nº 1425	April 1907	Nº 1148	Nº 4425	Nº 2855	-	October 1916	October 1916	-	-	-		1st January 1947	-	Darlington
Nº 1426	May 1907	Nº 1151	Nº 4426	(Nº 2856)	-	June 1924	July 1926	-	-	-	King's Cross	29th February 1944	-	Doncaster
Nº 1427	May 1907	Nº 1149	Nº 4427	(Nº 2857)	-	November 1919	Slide Valves until w/drawn	-	-	-	-	20th February 1945	-	Doncaster
Nº 1428	May 1907	Nº 1150	Nº 4428	(Nº 2858)	-	July 1913	May 1942	-	-	-	-	15th June 1946	-	Doncaster
Nº 1429	April 1907	Nº 1152	Nº 4429	Nº 2859	-	May 1919	November 1930	-	-	-	-	8th August 1947	-	Doncaster
Nº 1430	April 1907	Nº 1153	Nº 4430	(Nº 2860)	-	July 1913	December 1935	-	-	-	-	3rd February 1944	-	Doncaster
Nº 1431	April 1907	Nº 1154	Nº 4431	(Nº 2861)	-	July 1913	Slide Valves until w/drawn	-	-	-	-	21st October 1944	-	Doncaster
Nº 1432	November 1907	Nº 1171	Nº 4432	(Nº 2862)	-	November 1916	November 1916	-	-	-	-	26th July 1945	-	Doncaster
Nº 1433	November 1907	Nº 1172	Nº 4433	(Nº 2863)	-	January 1927	April 1932	-	-	-	-	2nd May 1946	-	Doncaster
Nº 1434	November 1907	Nº 1173	Nº 4434	(Nº 2864)	-	November 1912	Slide Valves until w/drawn	-	-	-	-	26th April 1945	-	Carried out Stationary Boiler duties at Don. Carriage Works as s/b Nº 809 from April 1945 to circa 1965.
Nº 1435	November 1907	Nº 1174	Nº 4435	(Nº 2865)	-	October 1913	May 1922	-	-	-	-	20th December 1945	-	Doncaster
Nº 1436	November 1907	Nº 1175	Nº 4436	Nº 2866	-	October 1915	October 1915	-	-	-	King's Cross	5th August 1947		Darlington
Nº 1437	March 1908	Nº 1186	Nº 4437	(Nº 2867)	-	February 1923	Slide Valves until w/drawn	-	-	-		11th March 1944		Carried out Stationary Boiler duties at Don. Carriage Works as s/b Nº 796 from March 1944 to circa 1965.
Nº 1438	March 1908	Nº 1187	Nº 4438	Nº 2868	-	June 1923	Slide Valves until w/drawn	-	-	-	-	8th August 1947		Doncaster
Nº 1439	March 1908	Nº 1188	Nº 4439	Nº 2869	-	September 1916	July 1927	-	-	-		2nd January 1947		Doncaster
Nº 1440	April 1908	Nº 1189	Nº 4440	Nº 2870	(Nº 62870)	July 1920	Slide Valves until w/drawn	-	-	-	King's Cross	14th February 1948	February 1948	Doncaster
Nº 1441	April 1908	Nº 1190	Nº 4441	Nº 2871	(Nº 62871)	June 1918	Slide Valves until w/drawn	-	-	-	-	21st May 1948	May 1948	Doncaster
Nº 1442	June 1908	Nº 1193	Nº 4442	Nº 2872	-	April 1914	November 1918	-	-	-	-	23rd January 1947	-	Doncaster
Nº 1443	May 1908	Nº 1191	Nº 4443	Nº 2873	-	March 1920	November 1934	-	-	-	-	2nd May 1946	-	Doncaster
Nº 1444	May 1908	Nº 1192	Nº 4444	(Nº 2874)	-	May 1917	July 1929	-	-	-	-	2nd October 1945	-	Carried out Stationary Boiler duties at Don. Carriage Works as s/b Nº 790 from October 1945 to circa 1965. Boiler, Nº 9459, was new in October 1944 and was one of the newest boilers to be redeployed for stationary boiler use.
Nº 1445	June 1908	Nº 1194	Nº 4445	Nº 2875	(Nº 62875)	April 1918	Slide Valves until w/drawn	-	-	-	-	22nd January 1949	January 1949	Doncaster
Nº 1446	June 1908	Nº 1195	Nº 4446	Nº 2876	(Nº 62876)	March 1922	Slide Valves until w/drawn	-	-	-	-	31st January 1948	January 1948	Doncaster
Nº 1447	June 1908	Nº 1196	Nº 4447	Nº 2877	(Nº 62877)	May 1914	May 1914	-	-	-	-	7th November 1949	April 1952	Carried out Stationary Boiler duties at Don. Wagon Works from February 1950 to July 1950. Then on same duty at Don. North Carriage Shed from July 1950 to April 1952. Cut up @ Don.
Nº 1448	July 1908	Nº 1197	Nº 4448	(Nº 2878)	-	March 1919	March 1920	-	-	-	-	15th June 1946	-	Doncaster
Nº 1449	August 1908	Nº 1200	Nº 4449	Nº 2879	-	May 1922	June 1940	-	-	-	-	4th July 1947	-	Darlington
Nº 1450	August 1908	Nº 1199	Nº 4450	(Nº 2880)	-	January 1920	January 1920	-	-	-	King's Cross	15th June 1946	-	Doncaster
Nº 1451	August 1908	Nº 1201	Nº 4451	Nº 2881	(Nº 62881)	March 1922	February 1937	-	-	-	-	29th April 1949	June 1949	Stratford
Nº 1452	August 1910	Nº 1276	Nº 4452	(Nº 2882)	-	From New	From New	-	-	-	-	27th February 1946	-	Doncaster
Nº 1453	September 1910	Nº 1277	Nº 4453	Nº 2883	-	From New	From New	-	-	-	-	27th July 1946	-	Doncaster
Nº 1454	September 1910	Nº 1278	Nº 4454	Nº 2884	-----	From New	From New	-	-	-	-	7th September 1946	-	Doncaster
Nº 1455	September 1910	Nº 1279	Nº 4455	Nº 2885	Nº 62885	From New	From New	-	-	-	-	23rd January 1950	May 1950	In for cut up @ Doncaster 24th January 1950
Nº 1456	October 1910	Nº 1280	Nº 4456	Nº 2886	-	From New	From New		-	-	-	5th August 1947	-	Darlington
Nº 1457	October 1910	Nº 1281	Nº 4457	(Nº 2887)	-	From New	From New	-	-	-	-	30th March 1946	-	Doncaster
Nº 1458	November 1910	Nº 1282	Nº 4458	Nº 2888	-	From New	From New	-	-	-	King's Cross	29th April 1947	-	Doncaster
Nº 1459	November 1910	Nº 1283	Nº 4459	(Nº 2889)	-	From New	From New	-	-	-	King's Cross	16th July 1943	-	Doncaster
Nº 1460	November 1910	Nº 1284	Nº 4460	(Nº 2890)	-	From New	From New	-	-	-	King's Cross	27th October 1945	-	Doncaster
Nº 1461	November 1910	Nº 1285	Nº 4461	(Nº 2891)	-	From New	From New	-	-	-	King's Cross	31st August 1945	-	Doncaster

Chapter 9

WITHDRAWAL, PRESERVATION AND STATIONARY BOILERS

Withdrawal Of The Small Atlantics

All of the GNR Atlantics entered the Grouping in 1923. The very first Atlantic to be withdrawn was in October 1924, when large-boiler Atlantic engine Nº 1300 succumbed, with four-cylinder compound Nº 3292 going next in January 1927. In 1935, soon after the introduction of the first of Gresley's A4 class 'Pacifics', Nº 3982, based at Hitchin and Nº 3988 of Peterborough were the next to go. The last main survivor to go was Nº 3252, which was withdrawn from service at Retford in July 1945.

The first of the class to be constructed, 3990, *Henry Oakley*, was withdrawn from service on 28 October 1937. Although it wasn't restored to running condition, it was restored externally to a GNR livery, with the red-brown frames omitted, which were painted black instead at Doncaster Works. However, Nº 990 isn't exactly in its original condition; in common with other members of its class, it acquired an extended smokebox. It left Doncaster on 23 November 1937, hauled north by J24 class engine Nº 1822 to York, prior to it being placed on 21 January

The very first tender Atlantic to be withdrawn from service was large boiler engine 1300. This 4-4-2 engine was a 4-cylinder compound that was built in July 1905 by the Vulcan Foundry. It was withdrawn in October 1924 from New England depot having been a highly unsuccessful engine. (Author's Collection)

Cross – reverse of outward route – Leeds Central.

Bill Hoole drove N° 251 and Alf Cartwright of Copley Hill, Leeds, drove N° 990. The maximum speed down Stoke Bank was 80mph. Ted Hailstone of King's Cross drove N° 60014, *Silver Link*, which was his regular A4 engine. The load was 410 tons. Peterborough was passed in seventy mins thirty-nine secs from King's Cross, despite 'permanent way' work at Potters Bar. 97mph was recorded at Three Counties and Biggleswade and a minimum speed of 70 mph was reached on Stoke Bank.

A similar trip was run a week later on the 27 September and was operated from King's Cross to Leeds with a stop at Doncaster, with the GNR veterans again hauling one leg of the trip. N° 251 steamed poorly on these trips, because the superheater had been removed, although the boiler flues had not been replaced with small tubes to compensate. In 1954, 251, worked more specials, but this time it was often helped by 4-4-0, Improved Director, D11 class engine N° 62663, *Prince Albert*.

After a visit to the Doncaster Paint Shop, 251 then entered the original York Railway Museum in March 1957. It was then transferred to the new National Railway Museum in 1975. It continues to be preserved in a static condition as a part of the National Collection.

Ivatt's large boiler
Atlantic 251 was built at Doncaster in 1902, was withdrawn from revenue earning service in 1947. It is seen here in the Main Hall at the National Railway Museum, York, as a valued exhibit of the National Collection.
(Author's Collection)

GNR small boiler
Atlantic 990, *Henry Oakley*, was built in May 1898 at Doncaster Works. In 1923, under the LNER's first renumbering scheme, it became 3990. After being withdrawn from service in October 1937 from Lincoln MPD, it was repainted in GNR livery and was preserved at the National Railway Museum, York, where it is seen here.
(Author's Collection)

Henry Oakley

Before entering the National Railway Museum in 1975, as a part of the National Collection, N° 990 was steamed again in 1975, in connection with the Stockton & Darlington Railway's 150th Anniversary Celebrations. During the summer of 1977, *Henry Oakley* was taken from the museum, under haulage of BR Standard 9F class 2-10-0 N° 92220, *Evening Star*, to Keighley to be loaned to the Keighley & Worth Valley Railway to work passenger trains. On 1 June 1977, *Henry Oakley* worked a train including Pullman Car *Mary*, named after Bishop Eric Treacy's widow, between Keighley and Haworth. On 14 August 1977, it worked double-headed passenger trips with the LYR's 0-6-0 N° 52044. It also ran trips every second weekend of every month until October. *Henry Oakley* was exhibited at the 'Doncaster Works Rail 125' event on 17 and 18 June 1978, in company with GNR 'big brother' N° 251.

Together with N° 251, N° 990 is normally housed at the National Railway Museum, York.

Stationary Boilers

As is usual with all railway systems, every item of rolling stock is fully utilised before going for scrap and here was no exception. Many steam engines after being officially withdrawn from revenue-earning service continued their service to the railway industry by being used as stationary boilers. Here are some examples of Atlantic type engines that were used as stationary boilers:

Withdrawn from service on 21 August 1947, large boiler Atlantic C1 class N° 2812 was used at Doncaster Wagon Works as a stationary boiler from September 1947-June 1952.

Withdrawn from service on 2 July 1948, large boiler Atlantic C1 class N° 2821 was used at Doncaster Carriage Works as a stationary boiler from July 1948-August 1950.

Withdrawn from service on 22 July 1948, large boiler Atlantic C1 class N° 2829 was used at Doncaster Carriage Works as a stationary boiler from July 1948-July 1953.

Withdrawn from service on 19 July 1948, large boiler Atlantic C1 class N° 2849 was used at Doncaster Carriage Works as a stationary boiler from July 1948-July 1953.

Withdrawn from service on 7 November 1949, large boiler Atlantic C1 class N° 2877 was used at Doncaster Carriage Works as a stationary boiler from February 1950-July 1950, then at Doncaster North Carriage Shed until April 1952.

Withdrawn from service on 2 April 1943, small boiler Atlantic C2 class

Seen in November 1950, joined together like 'siamese-twins' to effectively form a single boiler, are Ivatt Atlantics N° 3274 and N° 3285. They were placed adjacent to Doncaster Works Crimpsall Shop for a number of years after the war and were finacially cut up in March 1952. (Neville Stead Collection)

N° 3254 was used at Doncaster Carriage Works as a stationary boiler from April 1943-1964.

Withdrawn from service on 2 May 1946, large boiler Atlantic C1 class N° 3274 and C1 class N° 3285 were retained as complete engines and stood adjacent to Doncaster Works Crimpsall Shop for a number of years after the war as stationary boilers. They were joined together like 'Siamese twins' by both steam and water pipes, to effectively form a single boiler from September 1946 and were noted on 26 November 1950 as still in use. They were both cut up in March 1952.

Withdrawn from service on 26 July 1945, large boiler Atlantic C1 class N° 3278's boiler was used at Maldon Sawmill circa 1948 until 1975. This boiler was scrapped at Brightlingsea c1988-89.

Withdrawn from service on 6 October 1945, large boiler Atlantic C1 class N° 3287's boiler (N° 9359) was moved with C1 class N° 3278's boiler (see above) from Doncaster to Maldon in the late 1940s and is now being restored to working order at the Bluebell Railway. The boiler, which was new in August 1943, was used at Maldon Sawmill circa 1948 until 1975.

Withdrawn from service on 7 December 1945, large boiler Atlantic C1 class N° 3296 was used at Doncaster Carriage Works as stationary boiler N° 798, from December 1945 until c1965.

Withdrawn from service on 4 January 1946, large boiler Atlantic C1 class N° 4413 was used at Doncaster Carriage Works as stationary boiler N° 810, from January 1947 until c1965.

A photograph of a booster fitted large boiler Atlantic C1 class, N° 4419, was published in *Meccano Magazine* in the late 1950s, in

A row of large boiler Ivatt Atlantics is seen at the back of Doncaster Carriage Works on 17 May 1953. Left to right are engine 4444, stationary boiler 790; 3296, stationary boiler 798; 4437, stationary boiler 796; 4434, stationary boiler 809 and 4413, stationary boiler 810. (Neville Stead)

stationary boiler duty at Doncaster Works. It was built at Doncaster Works to works N° 1117, in 1906 and became GNR N° 1419, before being renumbered as N° 2849 in 1946. It was withdrawn from Ardsley on 19 July 1948 before receiving its BR allocated number of N° 62849. But after its withdrawal, it was then put to good use as a stationary boiler at Doncaster Carriage Works from July 1948, until July 1953, after which it was then scrapped at the nearby Doncaster Works. In 1927, it achieved some notoriety when it was sent to Scotland for trials over the Waverley route. However, it deemed itself insufficiently powerful enough to keep to time on the gradients involved and so it returned to England.

Withdrawn from service on 26 April 1945, large boiler Atlantic C1 class N° 4434 was numbered as stationary boiler N° 809 and was used outside Doncaster Carriage Works until c1965.

Withdrawn from service on 11 March 1944, large boiler Atlantic C1 class N° 4437 was numbered as stationary boiler N° 796 and was used outside Doncaster Carriage Works until c1965.

Withdrawn from service on 2 October 1945, large boiler Atlantic C1 class N° 4444 was used at Doncaster Carriage Works as stationary boiler N° 790, when it was withdrawn in October 1945, until c1965. This boiler, N° 9459, was brand new in October 1944 and was therefore one of the newest

This picture was taken at Doncaster Works during 1964 and shows a former C1 class Atlantic engine's frame onto which is mounted a Pacific type boiler. (Barry Collins)

boilers ever to be redeployed for stationary boiler use.

Noted at Doncaster during 1964 were the frames of a former large boiler Atlantic C1 class engine, which had mounted on them an A1/A2 class Pacific-style boiler, of diagram N° 117.

Two C1 class Atlantic boilers were in use at Bradford City Road Goods Yard in the mid-1950s, providing steam for a pair of compound hydraulic pumping engines.

Four stationary boilers were discovered in 1986 by Engine Shed Society members Nick Pigott and Steve Dymond at Maldon, Essex, where they had been used for powering sawmills in a timber yard. Two boilers were from former GNR C1 class Atlantics. One of these boilers now forms the basis of the Bluebell Railways Brighton Atlantic project, however the other former GNR boiler was scrapped at Brightlingsea in the late 1980s. One of the two smaller boilers was believed to be from former M&GNR engine 50 and is now based at the Mangapps Farm Railway, in Essex. The other boiler is thought to be from a former LSWR T10 class 0-4-4T engine and is now at the Avon Valley Railway. This image was taken at Maldon, Essex, in 1986 and shows the condition that the four boilers were in when they were discovered. (Michael J. Collins)

Chapter 10

EPILOGUE

by R.H.N. Hardy
Former Assistant District Motive Power Superintendent at Stratford Depot

Richard H.N. Hardy was born in 1923 and was educated at Marlborough College, as indeed was Sir Nigel Gresley. Richard served an apprenticeship at Doncaster Works and Carr Loco shed between 1941 and 1944. He then spent a further thirty-eight years working in the railway industry. He served in East Anglia and was Shed Master at Woodford Halse from the autumn of 1949, after which he moved to Stewarts Lane Depot and became the Shed Master there between August 1952 to January 1955. He then moved on to the Stratford District in general and Stratford Depot in particular as the Assistant District Motive Power Superintendent. SNCF Experiences on Nord and Est regions were obtained between 1958-71 and one of his last jobs was Divisional Manager in Liverpool on the London Midland Region. He retired in 1982 and is the author of *Steam in the Blood*, 1971; *Railways in the Blood*, 1985; and *Beeching: Champion of the Railway?*, 1989. He has also written numerous articles about railways and the Introduction to *The Railwayman's Pocket Book*.

When the twenty-two Klondikes started life in 1898, they were the largest passenger engines operating in the country. The first example, 990, was the only one to carry a name – *Henry Oakley*. It is seen here as it first appeared in service without its name.
(Peter N. Townend Collection)

Here are Richard's thoughts about the Great Northern Railway's Atlantics.

The Great Northern Atlantics

Firstly, the 'Klondikes' and they started life in 1898, the first one being N° 990, the only one to carry a name, *Henry Oakley*. They were excellent engines and by the standard of the times they were far from small, but not to be compared with the large Atlantic N° 251 that came out in December 1902. There were twenty-two 'Klondykes' including N° 271, of which more anon and the GN Atlantics when they came out were the largest passenger engines in the country and for a start their performance didn't quite measure up to their impressive appearance. But during the next thirty years starting in 1910, practical common sense prevailed, technical improvements were made and those sound if slightly disappointing engines gradually became members of a class of world beaters. What thrilling old things they were and what pleasure it has given me to remember that I served them in their wartime prime.

I was seven and a half when, in April 1931, I went to stay at Mexboro' with dear friends: an industrial town if ever there was one where the railway and the collieries were predominant. I was passionately interested in steam locomotives even at that age and we had come from London, so we changed at Doncaster into a train which was going to Barnsley, hauled by an old Great Central D7. And then while we stood there and I was gazing at the scene, in comes very gently into number 3 Platform, a 'Klondyke'. I can see it now and it was *Henry Oakley* herself, on the Hull-Liverpool express going through to Sheffield, and then home with stopping trains to Mexborough via Penistone and Barnsley.

Ten years later I was in Mexboro' again and, by then, I had the makings of a railwayman. I went to see my friends, Charlie and Emily Hepworth. Charlie had been a fireman in 1933 on the Hull and York jobs, he was settled on an absorbing railway career on the footplate and he was an excellent writer on technical matters in the ASLEF Locomotive Journal. All seemed set for him to become a driver as he had just taken and passed the examinations with flying colours when, on his way to work, he was hit in the eye by a snowball. That in fact finished his footplate career as he could no longer pass the eyesight examinations. But he continued his writing for his knowledge was profound and he had various jobs in the 'Loco' at both Mexboro' and Wath. However like any other true 'Poggy' man – a man who worked on the Great Central, he had no room for old GN 'knock-johns' until the 'Klondykes' came to work the Sheffield, Hull and Cleethorpes jobs. But in no time, they had those 'dyed in the wool' GC men in the palm of their hands. Charlie had the highest opinion of them at a time when anything else that came out of Doncaster was anathema at Mexboro'. But from the time we first met in 1941, he would always tell me to get to know everything I could about the famous large boiler Atlantics.

In fact there were only twenty-one true 'Klondykes' and the exception was N° 271, which was built as a four cylinder engine, with 15" x 20" cylinders. It was a pretty useless article lucky to last nine years before it was comprehensively rebuilt, after which she had two inside cylinders, a superheater and the same front-end as the last engines of the Ivatt regime and the early part of Gresley's. But whereas one had to scheme for steam with those very economical D1, J2, J6 and N2s, N° 271 would steam on a 'fag paper'. Even so, she had gone four years before I joined the LNER railway in 1941 and I would have

The exception to the rule with the general Klondike design was 271, which was built as a four-cylinder engine. It was lucky to last nine years before it was comprehensively rebuilt. 271, is seen here in its rebuilt form when Richard Hardy said that '...N° 271, would steam on a fag-paper'. (Peter N. Townend Collection)

left it at that, but it so happened that I was working in D shop in Doncaster Plant Works with a certain Eddie Maleham on his shaping machine. We got on very well together and he pointed one day to a middle-aged man who was operating with tremendous vigour a large machine across the gangway. This was the extremely outspoken and fierce Paddy Finnerty, but Eddie told me to go across and ask him to tell me about his days as a fireman at the Carr Loco. And it turned out that Paddy had had N° 271 as his own engine in one of the Atlantic links. But in the later 1920s, promotion for footplate staff had come to a stand and he left Doncaster to become a locomotive inspector on the Egyptian State Railways. I have no doubt he gave everybody hell in Cairo and he transmitted some of that fierceness to me until N° 271 was mentioned but, when he began to talk about her, he was a different man. He spoke quietly and thoughtfully about her, his unbeatable regular engine and described the journeys that he made to Peterboro' Nottingham, Hull, Leeds and so on, all but the top link London work. His whole attitude to railways and his work had changed and his torrid language was silenced. He had returned to Doncaster when the war started and was too old to go back to the footplate, but the Plant was glad of his services and he stayed there until he retired and I believe he lived to a great age.

In 1953, as everybody knows, N° 990 along with the first large Atlantic N° 251 worked a very special train to Doncaster. That was a great event and I had been invited to travel on N° 990 to Peterborough one evening while the two engines were finding their feet, so to speak, on the 17:00 from King's Cross to Peterborough. This had been arranged by the driver, my old Bradford mentor who had transferred to King's Cross in 1950 – Ted Hailstone. He had made the arrangements with the Inspector, George Wickens so I had not applied for a pass to ride on the engine.

However, on the platform was my old chief, Mr L.P. Parker, who cornered me until the last moment and bombarded me in his usual penetrating way with questions about my life on the Southern at Stewarts Lane Depot. It certainly would not do to jump up onto N° 990 in front of the Motive Power Superintendent, so I hopped in the front brake van and changed over later on at a signal stop at Holloway to come to terms with N° 990. We had a splendid trip and the four of us shared the work when we returned with a van train to King's Cross.

That old engine rode perfectly and used as I was to the large Atlantics which were wild-riding to say the least, the old 'Klondike' just slid along as Ted had her well notched up with the regulator fully open.

When I knew them, the GN Atlantics – the large ones – were in their final form and despite their discomforts were without doubt all-time greats. The first was N° 251 on which a number of experiments were carried out and put in place on N° 272 after which a total of ninety-four engines had been turned out at Doncaster Works by November 1910.

After the last war and as the B1s arrived in profusion, the Atlantics were withdrawn from service until 1950, when the last one was taken out of traffic. N° 4428 was the last engine to receive piston valves in place of the Richardson balanced valves and this left twenty-four with slide valves. It meant that in theory they were less powerful than those with piston valves which had 20" cylinders. I only knew them between 1941 and 1945 and then it was the Copley Hill trio N° 3280, N° 3300 and N° 4433, and the Sheffield and Grantham engines that I knew best.

The major improvement to the Atlantics was not so much to replace the slide valves but gradually to increase the influence of the superheater from 18 elements to 32 which produced steam, which was put to very good use at something like 400 degrees centigrade which is very hot indeed. Gresley initiated this gradual change which also included the use of his own twin-tube pattern, but in the end every member of the class had this very effective 32 element Robinson superheater.

Here is an example of what could be done with a piston valve C1 class, large boiler Atlantic and it was a miracle that Cecil J. Allen was on the train en route from King's Cross to Darlington for the engine and crew achieved work, far, far beyond what was theoretically impossible! An A3 class Pacific N° 2595, Trigo, had come down from King's Cross with the seventeen-coach 13:50 to Newcastle and which

When A3 class Pacific 2595, Trigo, worked down from King's Cross with the seventeen-coach 13:50 service to Newcastle, it failed at Grantham with a serious defect. In the absence of a replacement Pacific, large boiler Atlantic 4404 was the only engine available. So using a Gateshead crew with little if any experience of a GN Atlantic, they set off and gave 4404 plenty of stick and kept bowling along. It worked through to Newcastle with an assistant class V Atlantic and gained six minutes on a Pacific schedule from Newark to Selby. What an engine and what men! Here it is seen taking water, working a prestigious Pullman service on a different working. (Author's Collection)

had failed at Grantham with a serious defect. N° 4404 was the only available engine, well-stricken in miles since her last General but, in the absence of a Pacific, the best on hand. The crew were Gateshead men with little if any experience of a GN Atlantic but no doubt the advice they were given was common sense, give her plenty of stick and keep bowling it in and keep the fire, steam and water as high as is practical. And this is exactly what they did and you can bet that any onlooker could see her coming miles away. But that big fire was white hot and the Geordie fireman Barrick kept everything just so from start to finish. Brilliant! Cecil J. Allen – CJA is really well worth quoting considering that the well-filled train weighed an estimated 585 tons next stop York in ninety minutes for 82.7 miles. Furthermore, the enginemen had little or no experience of the Atlantic but plenty with A1s and A3s, which would serve the fireman well and the driving which had to be sensibly ruthless.

After a considerable struggle to get moving on the curve, they gradually worked up to 74mph below Claypole and CJA sat up and took notice. Three times on the journey to York, he rose from his seat to ensure that they had not left any coaches at Grantham for he simply could not believe what was happening. Having gone over Markham summit at 48mph, they then covered the 17.4 miles from Retford to Doncaster in fifteen minutes four seconds sustaining 72-75mph across the gently falling grades towards Scooby troughs, uphill past Bawtry and 61 at the top of the 1 in 200 at Pipers Wood. Quite incredible and had it not been for signals on before York and the driver being called down by hand well beyond the platform, they would have arrived in eighty-six minutes. N° 4404 which worked through to Newcastle with an assistant class V Atlantic, had gained six minutes on a Pacific schedule from Newark to Selby, a feat that CJA would never have believed possible had he not been there. What an engine and what men!

Photographs of Atlantics are so common that the unusual stands out clearly. There is, however, a well-known photograph of N° 4446 passing Potters Bar in the snow at the head of the 'Silver Jubilee' express. She had come up from Peterborough and held her own on the schedule of the fastest train in the country and she was a slide valve engine! She must have made a first time start from Peterborough: the one trouble that enginemen experienced was reversing a slide valve engine if it refused to start the train from a station. Once on the move, there was little difference, but getting away from the station could be the very devil. I found this out in September 1941 when a school friend of mine, Dick Lawrence, who was an apprentice on the LMS at Derby, joined me at Sheffield Victoria. Our idea had been to travel on the 15:30 from Sheffield as far as Penistone where we would change for Barnsley and the West Riding. We climbed from

Seen at Grantham shed in September 1941 are Dick Hardy, leaning on the right, together with Basil de Longh, in the cab of a J4 class 0-6-0 of 1898 vintage.
(R.H.N. Hardy Collection)

Sheffield all the way going very well indeed behind a flat valve C1 class, N° 3296, and when we got to Penistone, we walked the few feet to the engine and immediately the driver asked us who we were and where we were going. My friend was from Derby LMS and I was from Doncaster LNER and immediately the driver said 'B------ Barnsley, you're coming to Manchester with me but only one at a time.' So up got Dick and away we went flat out round that canted curve and on towards the summit. But we were stopped at Hazlehead and told that there was wrong line working from Hazlehead to Dunford Bridge. N° 3296 raised no objection to setting back downhill but when she was put into fore-gear and the regulator opened wide, she refused to move, so Joe Oglesby, our driver, had to open the cylinder cocks to get rid of the steam in the steam chest, pull the reversing lever into back gear, give her steam and then go through the whole procedure with the hope that she would move the train forwards and away. Before she would do that, she was reversed quite a few times, which meant that Joe's arms had tugged or pushed that recalcitrant lever before he was able to thrash the old girl into submission. She made a brilliant climb of the rest of the bank and then we plunged into the Woodhead Tunnel and she began to rock and roll and kick and bounce and sway as the Atlantics loved to do. It was not dangerous in the slightest but Dick Lawrence not only lost his cap but nearly lost his nerve! Joe regained about eleven minutes going down towards London Road.

On the return journey, we reached Guide Bridge and Joe said as he was to say many times in the next few years 'Right, Dick, take her to Woodhead and don't spare my mate!' so there I was, my first trip on an 'Atlantic' doing the driving up the hill and I practiced all the things that I had heard about and the little problems I might face with the movement of the reversing gear. Although the lever tugged at my arms relentlessly, I was able to master it and had her set just right to attack the gradient and by the time we got to Woodhead and I

Ivatt designed large boiler Atlantic 4422 which is seen piloting a 2-6-2 V2 class engine at Grantham. It was allocated to King's Cross depot in June 1930 and was scrapped at Doncaster after it was withdrawn from service on 28 September 1946. (Author's Collection)

handed over to Joe, I felt on top of the world. Despite all this and after we had rocked and rolled down to Sheffield in the dark and with Dick who joined us at Penistone, Joe said what wonderful engines the Atlantics were. Most of the 'Grinders' detested GN engines and anything that came from the 'hands' of Ivatt or Gresley and of course they loved their own Great Central engines, but the C1s and the V2s were different and as good or even better than nearly everything that came out of Gorton. But wherever they went from King's Cross to York and Leeds, even to Cambridge, which had an allocation of four Atlantics, they were revered and at such depots as Hitchin which covered a fair amount of the Cambridge work, the enginemen had their own 'piston valvers'. I knew Frank Green who was a fireman at Hitchin and although the slide valve engine N° 3275 was not their own, occasionally they would get her and there was always trouble getting away from stations first time but they made up for it on the journey because she was so free running. Whoever designed the front-end of the Ivatt Atlantics did a wonderful job.

It was with the West Riding men – Copley Hill, Bradford and Ardsley – that I had over four years – a great deal of experience, on the steam locomotive and the drivers and firemen in that area went to endless trouble to help me gain deep practical experience and I was fortunate to meet Driver Bill Denman who was a truly remarkable engineman. He was also a delightful companion and had a great deal of fun with me usually at my 'expense'. Here is an example of what he could do: one horrible pitch dark evening in February 1942, he was working the Colchester-Leeds express forward from Doncaster over that stretch of railway always known as the 'West Riding' and much of it heavily graded. On top of that, we were booked to stop at South Elmsell and Hemsworth, the former a devilish place to restart with a heavy load. Almost invariably, the engine was one of the Copley Hill Atlantics, N° 3280, N° 3300 or N° 4433. On this occasion, he had N° 4433, which was a wonderful engine but her safety valves were blowing off quite hard at 155lb psi as against the normal pressure of 170lb psi.

They left Doncaster on time with fourteen 'buckeyes' – that is fourteen corridor coaches, packed to the roof largely with service personnel and this heavy train had to be started not only at Doncaster but also at South Elmsall and worst of all from Wakefield on the 1 in 100. Time to go and fireman Jim Edison quietly gives his mate the 'right-away', Bill opens the regulator wide and nothing happens – no forward movement. To start a train on that gradient it was necessary to ensure that the setting of the invisible cranks was such that the maximum amount of steam possible would pass to the cylinders. Otherwise there would be no movement and Bill reversed five more times to try to find the best position, all without success, and the train had moved back along the platform a coach length. Many men would have sent for assistance but Bill knew it could be done with patience and skill and next time, the old engine began to move very, very slowly forward. There was a whisper of exhaust from her chimney, but the forward movement persisted and then came a second exhaust and a third and then a fourth which was a real glorious Atlantic bark up the chimney and they were on their way. Not a word had been spoken, no panic, no blasphemy, just Bill gently smiling and quietly working out the best position to make her move to get the train away. So he worked the engine flat out all the way up the steep grade to Ardsley and she and Jim Edison answered the call magnificently. Then he let fly down the other side towards Leeds, the engine riding ever wilder and wilder and Jim called across to his mate 'For God's sake, ease up Bill, we shall all be killed.' And Bill turned to his mate with his lovely smile 'Don't worry Jim, I'll be with you.'

Now here are a few asides to finish. I have quite clearly stated the case for these wonderful engines but there were people who hated the sight of them which may be news to you. The artisan staff at the running sheds had no complaints for they were in fact easy to maintain except when you had to lift various components such as Mr Richardson's slide valves which Edgar and I had to manipulate. No, when I was working with Edgar Joe Elvidge in '2-bay' in the Crimpsall, he and some of the others in Bill Umpleby's gang cursed the arrival of an Atlantic or a K2 class on the pit. It was impossible to make a

Given works number 1116 and running number 1418, this engine was completed in May 1906. It's seen running at high speed with an express passenger service through the middle road, with its left-hand injector doing its job. (Author's Collection)

This is the scene of the 'Sheffield Pullman' service, being worked by large boiler Atlantic 4426, which was allocated to King's Cross depot in June 1930. (Author's Collection)

Standing in front of 4468, *Mallard*, at Doncaster Works are several apprentices. On the extreme left is Peter Townend – who wrote the Foreward to this book – and R.H.N. Hardy, standing on the extreme right and who wrote the Epilogue, when they were both apprentices. The introduction of Gresley's A4 class Pacific engines caused the first of Ivatt's small boiler Atlantics to be withdrawn from service in 1935.
(Peter N. Townend Collection)

'profit' on piecework with at least two classes of engine and the C1 was the worst. Edgar Joe used to rant and rave about this and also about the fact that our Chargeman, Bill Umpleby was scrupulous with his book work whereas on the pit opposite, 'Beef' Taylor was said to be 'good with a pencil'. So life with Edgar was not one of slippered ease but we had a wonderful eleven months together during which time he called me every conceivable name under the sun and the more he swore at me the more I laughed.

Taking water at speed goes back to the 1860s but it was not what one might call an exact science although it was a simple matter to fill the tank but not always to cope with any overflow! The Great Eastern engines such as the D16s and B12s used compressed air from the Westinghouse brake system to lower the scoop to pick up water and to raise it clear from troughs that might still contain a fair amount of water. So long as the 'Westo' system was in good form, there was not the difficulty, but the Great Northern used the vacuum brake and things were very different on a Saturday afternoon in 1944, N° 4401 when a Grantham Atlantic, was at the head of twelve coaches leaving Doncaster, the rear six of which were ex-Carriage Works from somewhere down south. We were all stations to Grantham including even Barkston Junction, the driver was Bill Thompson and his mate Alf Rudkin whom I had known since the summer of 1941. Both men were real railwaymen and Alf who had done much to develop my abilities on both Pacifics and Atlantics, retired to the comfort of the train. Bill and I worked in perfect harmony and when we were nearing Muskham, my driver, standing tranquilly in his corner with his pockets as always stuffed with fruit, said: 'Right, Richard, we'll fill up at the "trawvs" and don't put the scoop down until I tell you.'

Now, the water scoops on the GN tenders were lowered by a long lever which the fireman, facing the tender, pulled towards him through an angle of about 45 degrees from the vertical. Primitive and very different to the little brass cock above the fire-hole doors on a B12 operated by a flick of the fingers while sitting down! On the GN, once the scoop was down and the water coursing upwards to fill the tank, the fireman did not have a snowball in hells chance of getting it out before reaching the end of the trough. And so I stood at the ready facing the tender and we must have been travelling about 50mph when we reached the troughs. And I waited and waited and waited and then without further consideration and youthfully thinking Bill had forgotten to tell me, I pulled at the lever down – far too soon. In no time, we were engulfed by a miniature 'Niagara', the footplate deep in coal-lumps, wet small stuff turned to sludge, up to the level of the fire-hole and me facing the long uphill grind towards Grantham. So I got busy tidying up, standing on coal , shovelling, sweeping while Bill's kindly eyes looked quizzically down at me above his splendid moustache: 'Never mind, Richard, you'll know better next time, have another pear!'

Here is a memory that will never die, a small thing but as alive today as it was early in 1942 as we ran slowly into Peterborough North. I was looking across towards the down side and there was a GN Atlantic facing south but clearly on its way home tender-first to New England. The fireman was standing in the time-honoured style with one arm resting on the cab cut-out and the other on the tender at the same comfortable level. And I suddenly realised that I was looking at a naturally dark-skinned fireman, a situation so rare in those days as to be quite remarkable It was Fireman 'Darkie' Dujohn, the son of a West Indian pedlar and well on his way to becoming a driver. A man much liked and treated by his fellow railwaymen as an equal and who, I have been told, died many years ago and relatively young. But I can always call to mind that the stance with his arms resting, the uniform cap pulled well down and the dark and cheerful face.

Leaving Doncaster in August 1945, I made my last journey to Leeds on the Copley Hill Atlantic, old N° 3280, one of the pre-war Pullman engines. I had come prepared to work and the driver was Walt Smith who I had only met once before. He welcomed me in the usual friendly West Riding way and told me that I was going to be the driver which was in fact the first time on one of the Leeds C1s. It was a lovely evening, we left about 18:30

On Dick Hardy's final day working 'on the footplate', the driver of A4 class Pacific 4489, *Dominion of Canada*, was Alf Cartwright. He always had something amusing to say and on this occasion he said to Dick, 'If this is your last trip with us, Dick, you had better be the driver to Donny; you've done enough shovelling to deserve it.' To mark the seventy-fifth anniversary of A4 class Pacific 4468, Mallard, breaking the world steam speed record, the National Railway Museum arranged for all six surviving A4 class engines from around the world to go on display at the National Railway Museum. Here is 4489, *Dominion of Canada*, on display during this very special event. (Author)

and stopped at every station, some of which have gone these many years, Hampole, Hemsworth, 'Fitzbilly' Halt near where a child named Geoffrey Boycott might well have been on his way to bed! It was a perfect fairly leisurely and happily conversational last trip but fate had not finished with me. I crossed over to the up platforms at Leeds Central and waited for the engine of the 'Mail' to back down. By that time engines were working through to London and vice versa and it was N° 4489, Dominion of Canada, a London engine. The driver was Alf Cartwright as usual wearing his celluloid collar: as always he had something amusing to say and on this occasion he said 'If this is your last trip with us, Dick, you had better be the driver to Donny; you've done enough shovelling to deserve it.'

Dick Hardy
Amersham, 2014

BIBLIOGRAPHY

Allen, Cecil J.: *British Atlantic Locomotive*, Ian Allan Publishing, Shepperton, 1968.

Groves, N.: *Great Northern Locomotive History ~ 1847-1866*, RCTS, 60 Hayse Hill, Windsor SL4 5SZ, 1986.

Longworth, Hugh: *British Railways Steam Locomotives ~ 1948-1968* Oxford Publishing Co., 2005.

Solomon, Brian: *Baldwin Locomotives*, Voyageur Press, 400 First Avenue North, Suite 400, Minneapolis, MN 55401, USA, 2010.

Solomon, Brian: *Baldwin Locomotives ~ Record of recent construction*, Schiffer Publishing Ltd., 4880 Lower Valley Road, Atglen, Pennsylvania 19310, USA, 2009.

Westing, Fred: *The LOCOMOTIVES that BALDWIN built ~ 1831-1923*, Superior Publishing, 1105 Tower Ave, Superior, WI, 54880, USA, 1966.

Yeadon, W.B.: *Yeadon's Register of LNER Locomotives*. Appendix One, named Engines, Book Law Publications, 328 Carlton Place, Nottingham, NG14 1JA, 2003.

Atlantic News ~ 'Reconstruction of 32424, BEACHY HEAD' Various issues

The Gresley Observer ~ 'The Journal of the Gresley Society' Various issues

The Railway Magazine Various issues

www.sixbellsjunction.co.uk